An Introduction to Aircraft Thermal Management

I0063494

An Introduction to Aircraft Thermal Management

MARK F. AHLERS

SAE INTERNATIONAL®

Warrendale, Pennsylvania, USA

400 Commonwealth Drive
Warrendale, PA 15096-0001 USA
E-mail: CustomerService@sae.org
Phone: 877-606-7323 (inside USA and Canada)
724-776-4970 (outside USA)
FAX: 724-776-0790

Copyright © 2020 SAE International. All rights reserved.

No part of this publication may be reproduced, stored in a retrieval system, or transmitted, in any form or by any means, electronic, mechanical, photocopying, recording, or otherwise, without the prior written permission of SAE International. For permission and licensing requests, contact SAE Permissions, 400 Commonwealth Drive, Warrendale, PA 15096-0001 USA; e-mail: copyright@sae.org; phone: 724-772-4028.

Library of Congress Catalog Number 2018951159
http://dx.doi.org/10.4271/9780768095524

Information contained in this work has been obtained by SAE International from sources believed to be reliable. However, neither SAE International nor its authors guarantee the accuracy or completeness of any information published herein and neither SAE International nor its authors shall be responsible for any errors, omissions, or damages arising out of use of this information. This work is published with the understanding that SAE International and its authors are supplying information, but are not attempting to render engineering or other professional services. If such services are required, the assistance of an appropriate professional should be sought.

ISBN-Print 978-0-7680-9342-1

To purchase bulk quantities, please contact: SAE Customer Service

E-mail: CustomerService@sae.org
Phone: 877-606-7323 (inside USA and Canada)
724-776-4970 (outside USA)
Fax: 724-776-0790

Visit the SAE International Bookstore at books.sae.org

Chief Product Officer
Frank Menchaca

Publisher
Sherry Dickinson Nigam

Director of Acquisitions
Monica Nogueira

Director of Content Management
Kelli Zilko

Production Associate
Erin Mendicino

Manufacturing Associate
Adam Goebel

contents

CHAPTER 3

Airplane-Generated Heat Sources   39

CHAPTER 4

External Heat Sources   53

CHAPTER 7

Environmental Control Systems   81

CHAPTER 11

Testing   149

introduction

"When the weight of the paper equals the weight of the airplane, only then can you go flying."

–Donald Douglas, Senior [Douglas Aircraft CEO 1957 to 1967]

Thermal issues on commercial aircraft have received increased attention during the past decade. Concerns over higher heat dissipations from *more electric airplane* (MEA) architectures and increased use of temperature-sensitive composite materials reached the fringes of mainstream media with the highly publicized 2004 launch of the Boeing 787 Dreamliner airplane program. The critical importance of thermal design on commercial aircraft would later garner front-page coverage in 2013, after the United States (US) regulator, the *Federal Aviation Administration* (FAA), joined Japanese regulators in grounding their country's 787s. This was in response to lithium-ion battery fires occurring on two separate Japanese aircraft. Regulatory agencies from other countries soon followed suit leading to a global grounding of the worldwide 787 fleet.

The likely cause of this disruption was a short-circuit, overheating one of eight battery cells, which overheated adjacent cells, causing *thermal runaway*. In thermal runaway a temperature rise increases the energy release rate further raising temperatures until the fuel source or oxidizer is exhausted. Deficiencies in the battery thermal design were a major contributing factor to this event.

Although there were no injuries or danger to passenger safety, due to robust airplane systems performing as designed, the economic loss to the airlines, and battery and aircraft manufacturer was enormous. Airlines lost the revenue from 50 airplanes taken out of service for 3 months and faced potential scheduling disruptions. Meanwhile, the battery and aircraft manufacturer shouldered the cost of redesigning, testing, and certifying a new battery to reduce the risk of failure, and aircraft modifications to improve passenger safety in the event of future battery failures.

While the 787 battery failure raised the importance of robust thermal design with one airplane component, opportunities for thermal-driven failures from equipment, transport elements, and structure affect much of an aircraft. All equipment creating physical movement, from the rotating engine turbine blade providing thrust propelling the aircraft forward to the actuators raising the main landing gear, entails the creation of waste heat from conversion inefficiencies. It starts when chemical energy in the fuel is converted to propulsive, electrical, pneumatic, or hydraulic power, and continues as the power is transferred to the point of use with wires, ducts, or hydraulic fluid lines. In all cases, heat rejected to the surrounding airplane structure and internal ambient environment can negatively impact airplane design and performance. The movement of electrons in the avionic equipment has a similar effect on airplane heat loads, driving increased interest in *aircraft thermal management* (ATM) from proponents of MEA.

System failures, such as a burst pneumatic duct carrying high-pressure and - temperature air, which escalates the thermal threat to passenger safety, require efficient thermal designs to minimize weight and cost.

Anywhere a fuel, oxygen, and ignition source can occur, including fuel tanks, cargo compartments, engines, and electrical equipment bays, the threat of fire must be addressed as part of the ATM plan.

Even sitting on an airport tarmac under the midday summer sun provides a significant thermal threat to the safe and economic implementation of the latest material systems, which are more sensitive to high temperatures and *ultraviolet* degradation compared to metals.

A technical literature search on, "Aircraft Thermal Management," reveals numerous papers devoted to systems responsible for aircraft cooling. *Electrical/electronic* (EE) cooling systems for removing avionics waste heat from the airplane and *air-conditioning* (AC) packs for providing air used to pressurize and cool the cabin dominate search engine results. While these systems are arguably the most important factors in maintaining a comfortable cabin ambient environment, other thermal issues receive scant attention beyond an airplane program thermal team. The lack of a comprehensive reference material for this critical aspect of airplane design, which affects so many different systems and organizations on an airplane program (i.e., electrical, environmental controls, flight controls, fuels, hydraulics, operations, propulsion, reliability, safety, structures), was my motivation for writing this book.

An Introduction to Aircraft Thermal Management, therefore, goes beyond the traditional coverage of systems designed to move waste heat from an airplane and provide cooling to look at all aspects of thermal management on the modern aircraft. It aims to meet the needs of both the non-technical and technical reader. More detailed explanation of terms and concepts will help the less technically focused airplane program personnel, of all disciplines, understand how their work affects thermal management on an airplane program. This is combined with key engineering equations designed to help a thermal engineer understand the physics behind the issues they are responsible for addressing. Meanwhile, observations from an aerospace industry insider with an intimate knowledge of the biggest thermal issues related to commercial airplanes during the past 30 years, ensure a focus on the most important aspects of this field.

This topic is divided among twelve chapters as follows.

Chapter 1 provides an overview of why aircraft thermal management is increasingly important to the implementation of the aircraft technologies that reduce fuel burn and improve passenger comfort. A more detailed explanation of many of the items mentioned follows in the remaining chapters.

Chapter 2 discusses the process of choosing both the aircraft outside ambient operating envelope and temperature requirements for passengers, cargo, equipment, and aircraft structure.

Chapter 3 covers heat generated by aircraft occupants and systems, which must be removed from an aircraft to maintain an acceptable temperature environment for equipment, structure, and passengers.

Chapter 4 delves into the primary external heat sources an aircraft faces, the sun, aerodynamic heating (for supersonic aircraft), and lightning. The equation set provided for calculating the reflected and direct solar load was picked based on its ease of use and ability to match common solar heating design guidelines.

Chapter 5 discusses design options for transferring waste heat generated by aircraft systems to the two primary aircraft heat sinks, outside ambient air and fuel.

Chapter 6 delves into aircraft fires and system failure events, which can generate damaging high temperatures. It also covers design approaches for containing the damage following a failure event long enough to ensure *continued safe flight and landing* at the closest useable airport.

Chapter 7 discusses the *environmental control systems* (ECS) responsible for controlling the cabin environment and protecting aircraft structure and systems from the dangers to aircraft survival from outside ambient conditions (ice).

Chapter 8 provides an overview of the primary thermal design options starting with insulation materials, the most common approach, and ending with material modifications for spreading the heat along structure.

Chapter 9 introduces the mathematical concepts and equations behind thermal and fluid flow modeling used to support ATM. The references mentioned for fluid flow modeling provide a comprehensive listing of loss coefficient that are likely to the meet needs of analysts modeling any ECS.

Chapter 10 provides examples of the different types of software used to support ATM. It discusses the process of choosing a software package, as well as the different sources, from open source to *commercial off-the-shelf.*

Chapter 11 deals with testing, ranging from material property to airplane level testing, including instrumentation options.

Chapter 12 mentions the few ATM differences between military and commercial aircraft.

The Appendix includes a list of acronyms, units, symbols, and dimensionless groupings mentioned throughout the text. They are also repeated at the first use in each chapter, except for units too common to justify cluttering the text.

Since each chapter is meant to stand alone, issues or ideas are frequently mentioned in multiple locations, with the most thorough explanation referenced in parentheses (i.e., Chapter xx or Section x.x.x). Both the *British Imperial* (English) and *International System* (SI) of units are shown, excluding the few times when it would require changing a referenced equation.

Mark Ahlers
Retired Technical Fellow/Thermal Marshal
Boeing Commercial Airplanes

CHAPTER 1

Why Aircraft Thermal Management Matters

"Pay attention to the details."

—TA Wilson [Boeing CEO 1972–1988]

1.1 Introduction

Aircraft thermal management (ATM) defines the management of heat in an aircraft to meet the *temperature requirements* of the vehicle and its occupants. This primarily involves *removing heat* and *protecting* equipment, systems, and structure from heat sources that could raise their temperature beyond design limits. Higher heat input equals higher temperatures, which often negatively impact structure and systems. Although excessively low temperatures, due to winter stopovers in Siberia or extended high-altitude cruise, can also adversely impact aircraft design or operations, high temperatures tend to drive more of the structural and system design.

1.2 Temperature Requirements

Crew and passengers (for commercial transport aircraft) must be neither too hot nor too cold during airplane operations in an outside ambient environment which can vary by more than 180°F (100°C). Maintaining *thermal comfort* (Section 2.3.1), therefore, is critically important. Poor temperature control can lead to passenger and crew complaints for the airlines and reduced airplane sales for the manufacturer. In military aircraft, with pilots under the extreme strains of combat, the additional stresses of an uncomfortable thermal environment can reduce reaction times or the quality of their decisions.

©2020 SAE International

Cargo compartment ambient temperatures must stay above 32°F (0°C) (Section 2.3.2) to keep liquids in passenger luggage and pets from freezing. This can be a concern with no animals generating heat to offset heat transfer to a –25°F (–32°C) or less *outside ambient temperature* (OAT) on ground or during flight.

Maintaining aircraft electronics, equipment, and structure temperatures over a specified design range (Sections 2.3.3 and 2.5) is vital to ensuring adequate performance over a guaranteed life. Local ambient temperatures are increasingly important since small variations have a greater impact on the lifespan of much of the latest equipment, such as energy efficient *light-emitting diode* (LED) lights.

1.3 Removing Heat

Air-conditioning (AC) packs (Section 7.2.2) remove heat generated by compressing outside ambient air used to pressurize, cool, and heat the aircraft cabin. Accurately estimating heat generated from passengers, crew, equipment (Chapter 3), and the sun (Section 4.2.1), is vital to avoid under (or oversizing) the AC packs.

Electrical/electronic (EE) cooling systems (Section 7.4) remove heat generated within boxes of densely packed electronics to exhaust to the cabin or directly overboard. Overboard exhaust avoids having to upsize AC packs to handle the heat, at an increased weight, cost, and fuel burn. Understanding the cabin environment and equipment heat loads is critical to ensuring efficient design.

Meanwhile, heat generated by the hydraulic system (Section 3.3.1) and engine oil coolers must match the cooling capacity of the wing fuel heat sink (Section 5.4.1).

1.4 Protection

ATM also involves protecting structure and equipment from heat generators (Chapter 3) including

- Engine-mounted electrical power generators
- Power feeders (Section 3.5) transferring electricity generated at the engines to the EE bay
- Bleed-air ducts (Section 3.7.1) transferring pressurized high-temperature air from the engines to the AC packs in the pack bay
- Hydraulic lines (Section 3.3.1) flowing high-pressure and -temperature liquids powering actuators moving wing control surfaces, raising and lowering landing gears, or opening and closing their doors
- Externally mounted collision avoidance lights mounted on the aircraft external skin or electronics powering internally mounted cabin mood lighting (Section 3.4)

Some airplane structure, systems, and components require protection from direct impingement of 1800°F (982°C) flames (Section 6.2). Meanwhile, systems and structure next to a ruptured bleed-air duct (Section 6.3.1), spewing out 450°F (232°C) air, must continue performing their intended function. Even the sun (Section 4.2.1) provides a significant thermal threat to aircraft structure that an ATM plan must address.

1.5 **Increased Importance**

ATM has become increasingly important, and sometimes the limiting factor, in vehicle design improvements in both the commercial and military arenas, for multiple reasons including

- The widespread use of *composite materials*
- *Increased heat loads* from expanded avionics functionality
- *More electric aircraft* (MEA) system architectures

1.5.1 **Composite Materials**

A composite material comprises two or more different constituent materials combined to create a new material with more desirable properties, such as higher strength at a lower weight. In airplane applications, composites often mean carbon fibers combined with a matrix resin. Most of the composite material strength comes from very strong and light fibers. Aligning and locating fibers to precisely match the load requirements reduces the amount of unneeded material (saving more weight) compared to metallic structure. The extreme strength-to-weight ratio of carbon fibers make them the preferred (fiber) choice, leading to the term *carbon fiber-reinforced plastic* (CFRP). Plastic refers to the matrix (resin) holding the fibers together while protecting them from abrasion and transferring load through the structure.

1.5.1.1 **ADVANTAGES OVER METALS**

Composites provide a primary means for reducing airplane *weight* and *maintenance costs*, and improving *aerodynamic performance* and *passenger comfort*. Meanwhile, the future offers the hope for reductions in *manufacturing costs*.

Weight. Boeing reported that CFRPs and other composite materials reduced the 787 airframe and primary structural weight approximately by 20% compared to traditional aluminum designs [1].

Figure 1.1 shows advanced composites reaching 50% of the 787-8 material makeup. The latest 787 competitor, the Airbus A350, advertises a 53% composite structure, leading

FIGURE 1.1 787 Structural material makeup [3].

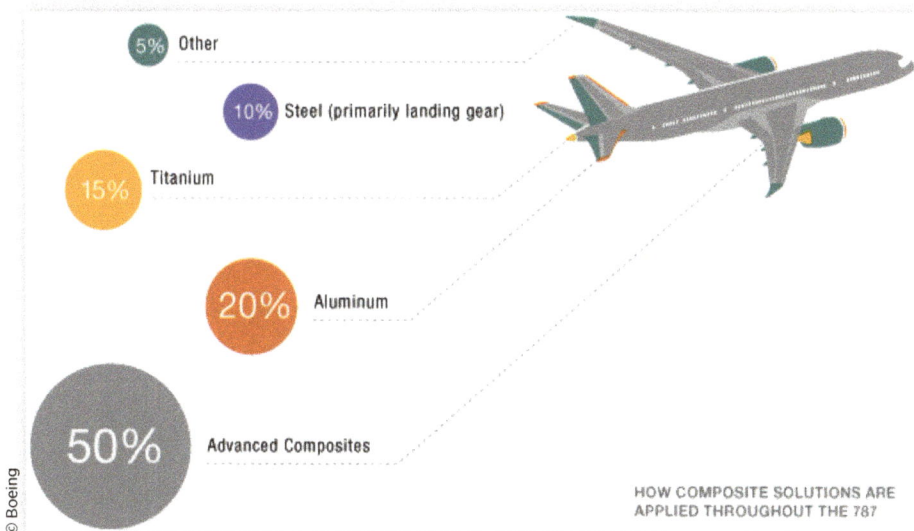

to a similar 20% weight reduction compared to a metallic airplane [2]. Weight is a primary driver of airplane fuel burn and a major reason for the expansion of composite material usage.

Composite structure can also lower airplane *maintenance* costs due to their reduced sensitivity to *fatigue loads* and a much lower risk for *corrosion*.

Maintenance. Aircraft operators must check airplane structure at defined flight cycles for signs of damage from wear due to fatigue stresses and corrosion.

Fatigue. Passenger jet aircraft fly at altitudes, typically around 35,000 ft (10,668 m), which avoid weather disturbances (for smoother flight) and support faster flight. The drag from higher density air occurring at lower altitudes limits aircraft speeds. Pressurized cabins make air travel possible at these higher altitudes, which lack sufficient *oxygen* (O_2) to support life.

Pressurizing the cabin causes tension and compression of structural loads during each flight as the structure expands ever so slightly (like a balloon). This generates fatigue stresses that damage the material, driving the need for costly maintenance inspections. Unpressurized wings face fatigue loads from in-flight bending causing tension and compression forces (Figure 1.2).

Corrosion. Bare metals react with O_2 in the air generating rust, which degrades the material and its properties in the process called corrosion. Liquid water or high humidity can accelerate corrosion.

Composite structure dramatically cuts fatigue loads and virtually eliminates corrosion, reducing inspection frequency and maintenance hours. Consequently, the Boeing 777 composite vertical tail requires 35% fewer maintenance hours than the smaller 767 aluminum tail [4].

Aerodynamics. A third advantage of composite materials is the ability to manufacture more complex structural shapes to improve aerodynamic performance. This primarily relates to making it possible to build longer thinner wings (Figure 1.2), providing more

FIGURE 1.2 787 Wing in flight.

1G Flight   ~12 ft

On-Ground   0 ft

1G Flight

787

© Boeing

lift and less drag, reducing fuel burn. The more complex curvature of each successive generation of composite wings further improves aerodynamic performance and fuel efficiency.

Comfort. Composite fuselages have led to more comfortable breathing environments from lower *cabin altitude* (higher cabin air pressure) designs. Cabin pressure refers to the internal ambient pressure corresponding to the cabin altitude, which indicates the amount of O_2 in the air.

The United States (US) *Federal Aviation Administration* (FAA) allows a maximum cabin altitude of 8000 ft (2438 m), based on studies showing a minimal health risk for most passengers from the reduced O_2 content compared to a sea level cabin [5]. Allowing the cabin pressure to rise above the sea level value reduces structural loads, which decreases aircraft weight, cost, and fuel burn.

The higher strength and reduced vulnerability to fatigue damage of composite structures reduce the penalties associated with maintaining a lower cabin altitude. Both Boeing and Airbus use this characteristic of composite fuselages to reduce the maximum cabin altitude on their 787 and A350 aircraft, respectively, to 6000 ft (1829 m). This lower cabin altitude is based on research showing a significant improvement in passenger comfort [6]. Decreasing the cabin altitude below 6000 ft, however, provides minimal increases in comfort.

Some manufacturers of composite airframes add humidifiers to increase the cabin air *relative humidity* (*RH*) and improve comfort, due to the reduced risk of corrosion. Outside air used to pressurize the cabin is extremely dry at cruise altitudes. While passengers add moisture to the air in breathing and perspiring, the cabin *RH* may be below 15% during cruise.

The *American Society of Heating, Refrigerating and Air-Conditioning Engineers* (ASHRAE) considers 30 to 60% to be a comfortable *RH* for office workers in sedentary environments. ASHRAE, an industry leader in the arts and sciences of heating, ventilating, AC, and refrigeration, sets many building codes. Commercial transport air quality and ventilation requirements are based to some degree on building standards, with modifications added to accommodate differences in an aircraft cabin.

Concerns over additional corrosion from more moisture condensing on metallic structure discouraged the addition of water to cabin air prior to the adoption of composite fuselages. With corrosion mostly eliminated, increasing the *RH* is more viable, although it increases the risk of moisture dripping on passengers or crew (in addition to adding weight and cost).

At cruise altitudes, moisture in cabin air can condense and freeze on aircraft skin and attached structure. During descent and landing, the ice melts as structure warms, generating water, which ideally flows unnoticed to bilge drains at the lowest point in the aircraft. While overhead crown designs include features to prevent condensed moisture from entering the cabin, some water may still drip on passengers. "Rain in the plane" is a 60-year-old problem for commercial transport operators [7].

Manufacturing Cost Hopes and Dreams. Although aerospace grade composite materials cost much more and face more challenging fabrication processes than metals, they offer hope for future savings from lower material and manufacturing costs, as the technologies mature [8, 9].

Shama and Simha [10] project the cost of composite fibers dropping in half from 2015 to 2020, due to expanded production, with manufacturing costs following a similar but less dramatic trend.

Manufacturing cost reductions are in addition to the existing financial benefits from a reduced part count. Careful tracking of each part during the life of the airplane, to support FAA certification, makes the airplane part count a major cost driver. The mountain of paperwork required to build and certify an aircraft led the late Douglas Aircraft CEO Donald Douglas to famously state (more than half a century ago), "When the weight of the paper equals the weight of the airplane, only then can you go flying." The composite material layup process, which combines multiple parts into larger continuous sections, contributed to a 700,000 part reduction on the 787 compared to the 767 [11].

1.5.1.2 DISADVANTAGES OF USING COMPOSITES

The many advantages of composites come with a few disadvantages that disproportionately affect ATM, such as

- Increased *strength sensitivity* to temperature changes
- *Lower thermal conductivity (k)*
- *Coefficient of thermal expansion (CTE) mismatches* with metallic structure
- Greater vulnerability to *lightning strikes*
- Higher risk and consequences of *analytical uncertainty*

Strength Sensitivity to Temperature. Most materials, including composites and metals, lose strength as their temperatures increase. Composite material strength tends to decrease at a faster rate, and at a lower temperature, compared to the metals they replace. The greater sensitivity (loss of strength at lower temperatures) can eliminate much of the potential weight benefit of a composite structure as material use increases to offset strength reductions at higher temperatures. Therefore, efficient thermal protection can have a larger effect on composite structural weight compared to a metallic airplane.

Lower Thermal Conductivity. The lower k of composite materials reduces conductive heat transfer, encouraging higher structure temperatures near heat sources, and reducing the fuel tank heat sink. Commercial and military aircraft dump system generated waste heat into their wing fuel tanks (Section 5.4). The lower k (higher thermal resistance) of composite wing panels resists ambient cooling of the fuel, reducing the fuel heat sink.

Lightning. A lightning bolt can carry over 4,740,000 Btu (5 billion J) of electrical energy into anything it strikes, including an airplane skin and attached structure [12].

Metallic airplanes can rapidly dissipate that energy over the entire fuselage or wing, usually avoiding significant damage or danger to airplane operations and passenger safety. Composite skins, unfortunately, have a much higher resistance to the flow of electricity. This prevents them from dissipating the large concentrations of energy rapidly enough to avoid localized high temperatures, which can permanently damage airplane structure or ignite fuel. Thermal designs (Section 8.12) must increase the electrical and thermal conductivity of the wing surface to safely dissipate these high energy levels.

CTE Differences. Materials expand and contract as their temperature changes, according to their CTE, which varies significantly between the mixture of metallic and composite structure aircraft contain. These differences in CTEs create thermal stresses in the airplane structure, after the airplane leaves the factory and enters an ambient environment which deviates from the factory.

Greater Analytical Uncertainty. Unlike metals, which have isotropic (constant) properties in all directions, composite material thermal properties have a strong directional dependence. Thermal conductivity is much higher along the fiber compared to the perpendicular direction. The properties also vary between material samples, leading to higher analytical uncertainty. The greater uncertainty of the material property associated with the greater impact of higher temperatures makes accurate structural thermal modeling both more challenging and a higher priority.

Awareness of the greater importance of accurate thermal assessment and design for composite structure drove the Boeing Company to create a program level position (a Thermal Marshal), specifically to address these types of ATM challenges.

1.5.2 Higher Heat Loads

Electrically generated heat loads on commercial aircraft have steadily grown with increased avionics functionality since their first use. *In-flight entertainment* (IFE) is a prime example of this trend. What started as a few projectors playing one or two movies half a century ago soon progressed to multiple *cathode ray tube* (CRT) television screens playing the same movies. Years later, individual *liquid crystal display* (LCD) screens, the size of an auto *global positioning satellite* (GPS) unit, provided each passenger multiple video and audio channels, bringing IFE into the modern electronics age. This eventually evolved into today's 13.3-inch LCD screens, offering 2,500 channels for A380 economy seating, which produces more waste heat than its humble predecessors [13].

A similar story exists for advanced weather radar, navigation, communication, and control systems, which all require tremendous computing power and generate increasing amounts of waste heat.

The military faces the added challenge of removing heat from advanced weapon systems and electronic countermeasures. An inability to remove the additional heat while maintaining aircraft stealth characteristics may limit future improvements in system capabilities. Ram air (Section 5.2.2.1), the primary means of removing heat on commercial airplanes, provides an easy target for heat-seeking missiles, limiting its use in military aircraft.

1.5.3 More Electric Aircraft

The upward trend in electrically generated power took a huge leap when it began replacing the major airplane system's power sources (hydraulic, pneumatic, and mechanical), leading to the *more electric aircraft* (MEA).

At service entry in 2011, the Boeing 787 Dreamliner (the first commercial MEA in terms of systems performance and control) brought with it an electrical generating capacity greater than 1.4 MW. This is four times more than its predecessor entering service about 30 years earlier, the 767 [14].

While some of the power increase was due to evolutionary improvements in weather radar, navigation, communication, and IFE; a much bigger factor was the development of a *no-bleed* systems architecture.

1.5.3.1 NO-BLEED SYSTEMS

Replacing bleed air extracted from the engines with pressurized air provided by electrically driven compressors consumes most of the additional 787 electrical power. The increasingly negative impact of bleed-air extraction on engine fuel burn and perfor-mance (Section 7.2.1) is a major driver for no-bleed systems. (Aircraft engines compress outside ambient air supplied to the combustor. Most jet aircraft bleed off some compressed air to meet aircraft system needs.)

On the 787, electrically driven systems replace those on past airplanes powered by pressurized air bled from engines. Electrically driven compressors provide high-pressure air to run the AC packs and pressurize the cabin. Electric resistance blankets prevent excess ice buildup on the wings. Engine-mounted generators provide power to start the engines, and *auxiliary power unit* (APU) generators and electric pumps pressurize the hydraulic system. On the 787, bleed air only provides engine inlet icing protection.

Removing the bleed-air system (Section 7.2.1.1) also saves on the weight and cost of heat exchangers (precoolers), valves and ducting, a leakage and overheat detection system, and challenging thermal design.

Electrically driven compressors offer additional energy savings by better matching the aircraft system pressure requirements compared to a bleed system. Outside ambient pressure swings, from 14.7 psi (101.3 kPa) at sea level to 2.35 psi (16.24 kPa) at 43,000 ft (13,106 m) (Section 2.2.1.1), creating a varying required air supply pressure to meet aircraft needs. A commercial transport jet engine bleed system can provide compressed air at two or three different pressures, which can far exceed the aircraft system needs. A variable speed electrically (or engine shaft) driven compressor, on the other hand, can provide the exact required pressure at all mission times.

1.5.3.2 ELECTRIC ACTUATORS

Electric actuators can replace most hydraulic system functions, such as moving aircraft control surfaces, landing gears and doors, or braking.

Potential benefits of electric actuation (compared to hydraulics) include eliminating the risks of a fluid catching fire, freezing, or leaking; higher efficiency; lower maintenance costs; easier installation; and improved *health monitoring*.

While hydraulic fluids developed for aviation have fire-resistant properties (making them difficult to ignite at room temperature), they can sustain combustion if heated above 356°F (180°C). This is far below the 1300°F (700°C) that brake temperatures can exceed following a high-speed rejected takeoff for a heavy aircraft [15].

Cold temperatures that cause hydraulic fluids to become more viscous, or even freeze (reducing actuator responsiveness to flight crew (or computer) commands) improve the current carrying capacity of electrical wires. Wire electrical resistance normally decreases with decreasing temperature.

Electrical wires are more reliable than 3000 to 5000 psi (143.6 to 239 kPa) hydraulic systems tubing and seals, which can leak or burst, expelling liquid over a large area. The impacted area from a wire failure, such as an electrical arc, is usually very limited, unless it ignites a flammable liquid creating an explosion.

Health monitoring, which involves checking performance characteristics for indications of impending failures, is easier to implement with electrically based systems compared to hydraulic ones. Performance data collected in-service helps determine component and system reliability and anticipate failures, which supports proactive parts replacement. Unscheduled maintenance delays, or cancelled flights from failed parts, are very costly for airlines.

Electrical actuators are, however, a poorer choice for areas with flammable liquids, where they create an ignition risk [16].

1.5.3.3 CHALLENGES

The biggest challenges to implementing MEA are weight, cost, and technical and schedule risk, in addition to waste heat removal.

Electronic boxes, which condition and filter the power, are heavier than initial predictions, and power equipment efficiencies have room for improvement. Like all newer technologies, their costs are high and past technical issues contributed to program delays in recent airplane programs. Aircraft manufacturers and systems suppliers will, therefore, likely trade traditional and more electric architectures for the foreseeable future.

References

1. Boeing Commercial Airplanes, "787 From the Ground Up," *Aero Magazine*, 4th quarter 2006, 2006, accessed January 3, 2017, http://www.boeing.com/commercial/aeromagazine/articles/qtr_4_06/article_04_2.html.

2. Marsh, G., "Airbus A350 XWB Update," November 16, 2010, accessed January 2, 2017, https://www.materialstoday.com/composite-applications/features/airbus-a350-xwb-update/.

3. Boeing Commercial Airplanes, 2006.

4. Ibid.

5. Muhm, J.M., Rock, P.B., McMullin, D.L., Jones, S.P. et al., "Effect of Aircraft-Cabin Altitude on Passenger Discomfort," *The New England Journal of Medicine* (July 5, 2007).

6. Ibid.

7. Peasant, T., "Rain in the Plane," *Aircraft Cabin Management* 2 (April 2013), accessed February 10, 2017, http://www.aircraftcabinmanagement.com/feature/condensation-in-aircraft?session_id=o6r49kjbtifvi02gp66563jj16.

8. Furst, K., "After Aerospace: The Future of Carbon Fiber Composites in the US," MarketResearch.com, September 13, 2016, accessed 1/20/2017, http://blog.marketresearch.com/after-aerospace-the-future-of-carbon-fiber-composites-in-the-us.

9. Grant, C., "Addressing the Cost of Aerospace Composites," CompositesWorld.com, 1/26/2011, accessed 1/20/2017, http://www.compositesworld.com/columns/addressing-the-cost-of-aerospace-composites.

10. Shama, R.N., Simha, T.G.A., Rao, K.P., and Ravi Kumar, G.V.V., *White Paper, Carbon Composites are Becoming Competitive and Cost Effective* (Bengaluru, India: Infosys, 2015), https://www.infosys.com/engineering-services/white-papers/Documents/carbon-composites-cost-effective.pdf.

11. Boeing, "World Class Supplier Quality—Boeing 787 Updates," accessed January 5, 2016, http://787updates.newairplane.com/787-Suppliers/World-Class-Supplier-Quality.

12. Institute of Physics, "A Bolt of Lightning Has Enough Energy to Toast 100,000 Slices of Bread," accessed January 5, 2016, physics.org. http://www.physics.org/facts/toast-power.asp.

13. Meacham, S., "Super-Jumbo Turns 10 Aviation's Big Bird is about to Turn 10: And Remains a Marvel of Engineering Excellence," *Sun Herald* (Australia), 09/03/2017.

14. Polek, G., "Boeing Will Reap Benefits of 787 Technology for a Generation," *Aviation International News*, November 12, 2011, accessed January 30, 2016, http://www.ainonline.com/aviation-news/aerospace/2011-11-12/boeing-will-reap-benefits-787-technology-generation.

15. Flight Safety Foundation, "Hydraulic Fluid as a Fire Source," Skybrary, Last modified September 21, 2017, accessed October 15, 2017, http://www.skybrary.aero/index.php/Hydraulic_Fluid_as_a_Fire_Source.

16. Gonzalez, C., "What's the Difference between Pneumatic, Hydraulic, and Electrical Actuators?" *Machine Design* (April 16, 2015), http://machinedesign.com/linear-motion/what-s-difference-between-pneumatic-hydraulic-and-electrical-actuators.

Temperature and Thermal-Related Requirements

"It doesn't matter what temperature the room is, it's always room temperature."

—**Steven Wright [American comedian]**

2.1 Introduction

Outside ambient conditions and the corresponding aircraft air, equipment, and structure temperature requirements (along with aircraft heat loads) drive thermal design. Increasing the maximum outside air temperature design requirement raises ambient heating of the cabin environment, while reducing the effectiveness of aircraft cooling systems using outside air as a heat sink (Section 5.2). Larger cooling system components that cost and weigh more and increase fuel burn, must offset losses in cooling system effectiveness. Decreasing the minimum allowable *outside ambient temperature* (OAT) for ground or flight operations can also negatively impact the economics of building and operating an airplane. Cold temperatures can drive the need for larger hydraulic lines to accommodate increases in fluid viscosity, restrictive fuel loading procedures to prevent freezing, and more fuselage skin insulation to ensure passenger *thermal comfort* (Section 2.3.1).

Choosing the appropriate outside ambient conditions to design to, therefore, is one of the most important tasks for vehicles facing variations of weather conditions across the globe. Aircraft designed for peak performance under the most extreme ambient conditions ever recorded cost more to build and operate, burn more fuel, and generate additional greenhouse gases. Limiting airplane operating capabilities to an insufficient temperature range, however, can cause too many cancelled flights for airlines used to the 99% dispatch reliabilities, reported by top Boeing 737 and Airbus A320 operators. Dispatch reliability is the percentage of flights an airplane dispatches within a specified time (usually 15 min) of their planned departure.

Designing for too narrow an ambient temperature range can be fatal for military aircraft if vehicle or weapons system performance is sufficiently degraded (during combat) to compromise the mission.

The process of finding the appropriate balance between airplane temperature capabilities and the customer needs starts with an understanding of the outside ambient conditions affecting aircraft performance and passenger comfort.

2.2 Outside Ambient Conditions

Outside ambient conditions affecting airplane temperatures include air, ground, and sky temperatures, wind speed, solar flux levels, air pressure, and *relative humidity* (*RH*). Ambient conditions chosen to demonstrate passenger airplane performance must be conservative enough to ensure *continued safe flight and landing* (CSFL) for all possible operating scenarios that can occur during the life of the airplane fleet. For the Boeing 777, entering service in 1995, *life* equals more than 18 million flight hours 20 years later from over 1,400 airplanes [1].

Ambient design assumptions must also be stringent enough to drive structure material selection and equipment designs that meet airplane performance guarantees. Historical *climate data* provide the starting point for determining the appropriate outside ambient design conditions.

2.2.1 Climate and Aircraft Performance

High ground OAT operations are a daily challenge for aircraft operations during summer months at airports throughout the Middle East and desert Southwest (in the United States [US]). One example is Kuwait City, which faces a 112°F (44°C) average high ambient temperature during the hottest month of the year and regular excursions beyond 120°F (49°C) [2]. Phoenix closely follows Kuwait City, with a 106°F (41°C) peak average high OAT [3], which seems chilly given increasingly frequent reports of record high temperatures.

While many frequent flyers can relate to the difficulty staying cool prior to takeoff from a packed airplane lacking adequate cooling during sweltering weather, few are aware of challenges faced at the opposite end of the weather spectrum. Extreme cold commonly occurs during multiple months of the year for airports in Siberia and Alaska, with ground OATs regularly dipping below −40°F (−40°C). Yakutsk, the capital of Sara in Siberia, Russia faces a January average daily low OAT of −43°F (−42°C), with a record −84°F (−64°C) reached one February day [4]. A population of 267,000 (2010) brings with it frequent large commercial aircraft service, providing sufficient sales to justify designing commercial aircraft to operate in this environment. Even lower ambient temperatures can occur during flight, since the OAT decreases with altitude due to a lower pressure causing reduced internal energy.

The *International Civil Aviation Organization* (ICAO) defines the rate at which outside air temperatures decrease, called the *environmental lapse rate*, as −6.5°C/km (3.56°F/ft) from sea level to 11,000 m (36,090 ft) [5]. The ICAO also assumes a constant air temperature between 11,000 m and 20,000 m (36,090 ft to 65,620 ft).

The actual lapse rate can also vary from the ICAO standard due to moisture in the air or weather patterns. Inversions, in which warmer air sits on top of much colder ground air, frequently occur on the coldest days from ground radiative heat transfer to the cold sky. Therefore, the typical −40°F (−40°C) Yakutsk winter ground OAT does not correspond to a −157°F (−105°C) overhead static air temperature at 33,800 ft (10,000 m).

$$-6.5°C/km*10km+-40°C=-105°C(-157°F)$$

Instead, the lowest recorded temperature at 10 km (6.21 mi) is a balmy (by comparison) −99°F (−73°C), according to the *MIL-HDBK-310 Climatic Data for Developing Military Products for US Military Vehicles* [6].

2.2.1.1 STANDARD DAY

The *International Organization for Standardization* (ISO) established the *International Standard Atmosphere* (ISA) or *standard day* for short, atmospheric temperatures as reference points for vehicle and equipment design and analysis (Figure 2.1).

FIGURE 2.1 Standard day atmospheric temperature and pressure change with altitude [7].

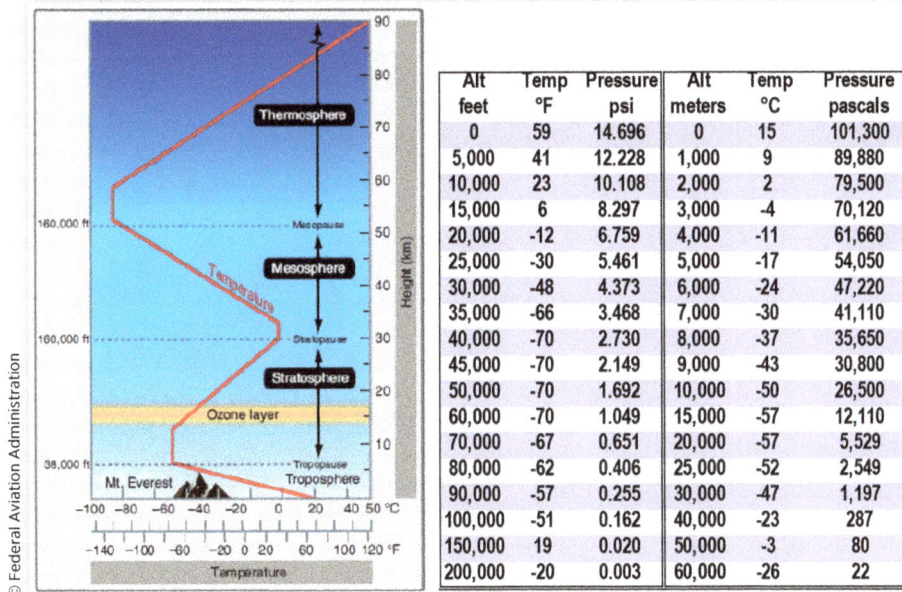

Alt feet	Temp °F	Pressure psi	Alt meters	Temp °C	Pressure pascals
0	59	14.696	0	15	101,300
5,000	41	12.228	1,000	9	89,880
10,000	23	10.108	2,000	2	79,500
15,000	6	8.297	3,000	-4	70,120
20,000	-12	6.759	4,000	-11	61,660
25,000	-30	5.461	5,000	-17	54,050
30,000	-48	4.373	6,000	-24	47,220
35,000	-66	3.468	7,000	-30	41,110
40,000	-70	2.730	8,000	-37	35,650
45,000	-70	2.149	9,000	-43	30,800
50,000	-70	1.692	10,000	-50	26,500
60,000	-70	1.049	15,000	-57	12,110
70,000	-67	0.651	20,000	-57	5,529
80,000	-62	0.406	25,000	-52	2,549
90,000	-57	0.255	30,000	-47	1,197
100,000	-51	0.162	40,000	-23	287
150,000	19	0.020	50,000	-3	80
200,000	-20	0.003	60,000	-26	22

© Federal Aviation Administration

While the standard day ground OAT is close to the average earth surface temperature, it does not represent it. Williams [8] sets the average earth surface temperature at 57°F (14°C), about 2°F (1°C) less than the ISA standard day values.

Airplane design conditions capture the hotter or colder ambient temperatures by adding a temperature difference to the standard day values, based on a probability of occurrence, as shown on an *environmental envelope*.

2.2.1.2 ENVIRONMENTAL ENVELOPE

Government regulators, like the US *Federal Aviation Administration* (FAA) and *European Union Aviation Safety Agency* (EASA), certify airplanes for *safe operations* over an environmental envelope defining OAT versus altitude (Figure 2.2).

Airplane manufacturers provide *performance guarantees* and *operational requirements* and *limits*, for flights occurring within the environmental envelope.

FIGURE 2.2 Typical aircraft environmental envelope (Adapted from [9]).

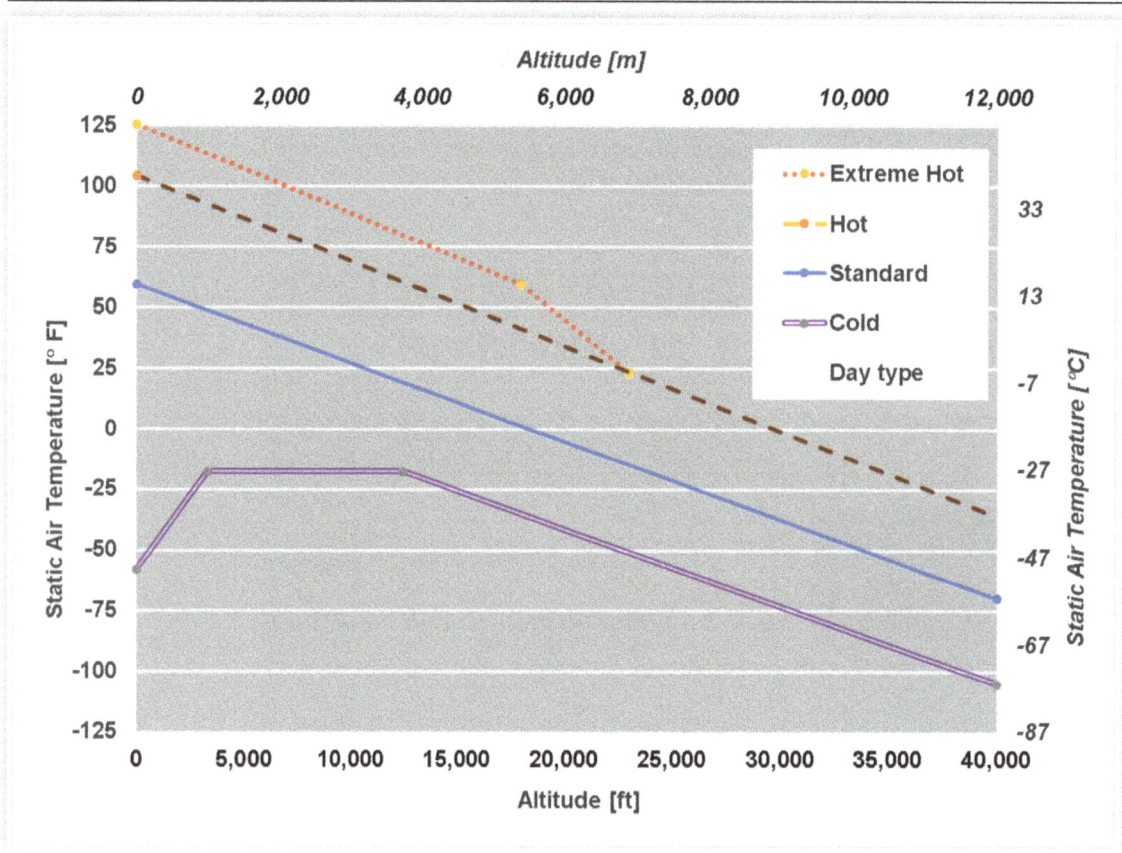

While Figure 2.2 shows temperatures plotted versus altitude, the more useful form for thermal analysts reviewing boundary conditions, flight manuals often show pressure altitude versus temperature.

2.2.1.3 SAFE OPERATIONS

Commercial transport airplane manufacturers must demonstrate safe operations by showing (through testing and analysis) that the chance of a catastrophic hull loss is extremely improbable. Extremely improbable means never occurring during the life of the aircraft fleet, based on the reliability analysis methods provided in the FAA Airworthiness Circular *AC 25.1309-1A*.

2.2.1.4 PERFORMANCE

Airplane manufacturers provide performance guarantees for the vehicle, systems, and individual components. Major vehicle level guarantees include fuel burn, maximum mission length, takeoff distance, rate of climb, stall speed, and engine power requirements and capabilities. At the system level, the maximum and minimum allowable cabin air temperatures during hot and cold OATs, and time required to cool down or warm up the passenger cabin drive ATM design.

2.2.1.5 OPERATIONAL REQUIREMENTS AND LIMITS

Operational limits include items like the maximum allowable airplane weight for takeoff during the hottest days, when the lower density of the outside ambient air reduces aircraft performance. Limits that matter to ATM personnel include the maximum OAT before the *electrical/electronic* (EE) cooling systems must be turned on, and the fueling procedures during extreme cold ground OATs (to prevent fuel in aircraft wings from getting too cold).

2.2.1.6 OAT AND AIRPLANE PERFORMANCE

OAT effects on airplane flight characteristics and engine performance drive the development of the ambient environment envelope more than the impact on the cabin environment. System designs may also accept degraded performance within this envelope for items not affecting passenger safety, such as cabin air temperature.

High OAT. Extremely high OATs, with their elevated summer solar fluxes, tend to size external structure for subsonic aircraft.

As the ground OAT increases, air density decreases, causing multiple negative consequences. It reduces the aerodynamic lift, which increases the airplane takeoff distance, minimum speed required to remain aloft, rate of climb, and required engine power. Lower density air also reduces engine performance as less mass flows through the engine inlet, providing less oxygen for combustion and less bypass air to push backward. Air bypassing the engine core, which the engine fan pushes backward, provides most of the thrust propelling an airplane forward (Section 7.2.1). The combined effect of needing more engine power as the engines become less efficient can force increases in the engine size and fuel burn. This makes concerns over the impact of higher OATs on airplane structure temperatures and cooling system performance seems minor in comparison. In the commercial airplane world, aerodynamics and propulsion system performance are the top priority since they have the greatest effect on fuel burn.

The two maximum temperature lines occurring below 18,000 ft (5490 m) (Figure 2.2) define varying requirements for different systems or airplane functions. Ambient scenarios causing the greatest aircraft system or structural impact can also include the interaction with factors beyond temperature. One example is an *air-conditioning* (AC) system, where a combined high ambient temperature and *RH* determines the design point. The extremely hot 122°F (50°C) ground outside ambient air is also bone dry. Meanwhile, the lower temperature (but more humid) outside air can take more energy to cool since moisture condensing in an airstream releases heat. This drives the need for a combined lower outside maximum hot temperature and high humidity requirement for the AC system.

Low OAT. While decreasing the OAT increases the airplane engine performance, safety concerns take over if it causes airplane fluid (fuel and hydraulics) or structure temperatures to drop dangerously low.

Frozen fuel or ice in the fuel lines can starve the engines. The viscosity of hydraulic fluids powering moveable airplane control surfaces also increases with decreasing temperatures, leading to a slower response to pilot or (more often) computer commands. Cold fuel can further cool hydraulic fluid when the system tubes run through the fuel tanks. Extreme cold temperatures can also increase *thermal stresses* from *coefficient of thermal expansion* (CTE) mismatches between composite and metal structure. Higher stresses can degrade the material system over time, but rarely pose the imminent threat to passenger safety of excessively cold fuel or hydraulic fluid temperatures.

Extreme cold temperatures can cause low passenger cabin temperatures with long warm-up times using average-sized ground cart heaters, while aircraft batteries quickly lose their charge. Operations in extremely cold regions may also incur special procedures to prevent damage to airplane systems during extended time on ground when heated hanger space is unavailable.

For economic reasons, the maximum and minimum temperature lines on Figure 2.2 fail to cover the range of possible OATs. Operating envelope design ambient temperatures are instead based on the probability of occurrences rather than the highest and lowest values ever recorded to limit overdesign. Airlines and military planners have learned to accept aircraft able to operate in environments covering all but a tiny percentage of the time as defined by historical weather data.

2.2.1.7 MIL-HDBK-310

US military aircraft/system designers use the extensive weather data published in the *MIL-HDBK-310*. Civilian aircraft manufacturers may also use *MIL-HDBK-310* or generate their own data tailored to commercial routes and operations.

MIL-HDBK-310 provides climatic data as a frequency of occurrence and extreme values, based on collected weather data and statistical analysis. The data covers regional locations and the whole world (Figure 2.3).

Figure 2.3 includes four regional land types based on their temperatures, *hot*, *basic*, *cold*, and *severe cold*, and a *coastal hot region* for ships.

Hot Region. The hottest temperatures, which usually receive the most attention in aircraft design, occur in the interior of northern Africa eastward toward India. This includes the Middle East, with its major commercial operators such as Emirates and Qatar Airlines, and all too frequent wars and military conflicts. Military and commercial operators face nearly identical climatic conditions.

Basic Region. The basic region includes a wide temperature range covering much of the earth's land surface, including the entire lower 48 US states, outside of a small area surrounding Death Valley, California. Basic region temperatures are available on both the hot and cold ends of the temperature spectrum, with and without high humidity levels.

FIGURE 2.3 Location of climatic region types of land areas of the world [10].

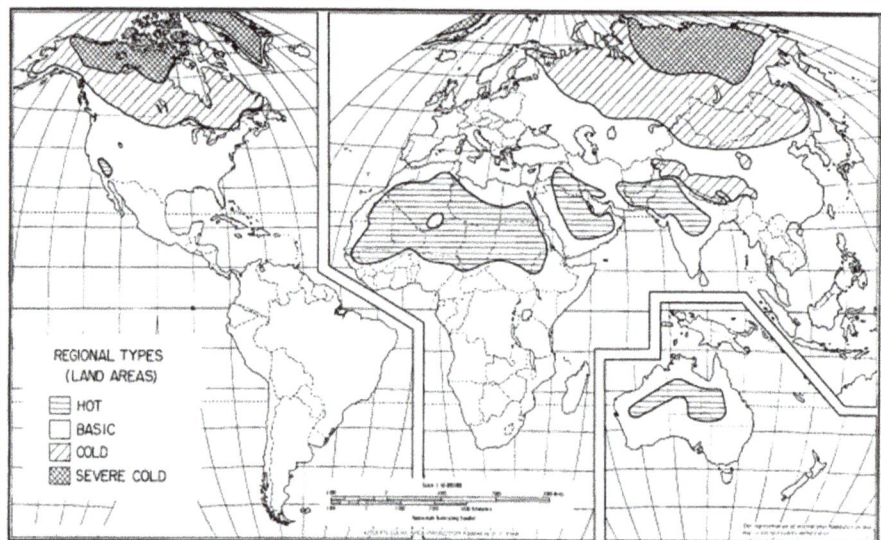

© Department of Defense

Cold Region. Operations in cold regions incur special procedures to prevent damage to airplane systems during extended time on ground when heated hangar space is unavailable.

Extreme Cold Region. A tiny percentage of worldwide commercial flights occur during the coldest days in the extreme cold regions of Alaska, Siberia, and Northern Canada. The primary concern with the coldest temperatures defined in these areas is avoiding permanent damage to airplane structure or systems rather than ensuring full performance capabilities. Commercial airplanes must be able to survive a diversion to an extreme cold airport, during an emergency, without compromising passenger safety or aircraft integrity.

Coastal Hot Region. *MIL-HDBK-310* includes the coastal hot region to support ship design. The combination of high temperature and humid air is also well suited for sizing aircraft AC packs (Section 7.2.2). Pack performance has the greatest impact on vitally important passenger and crew comfort.

Temperature Data. Table 2.1 shows diurnal temperatures for the 1-percent frequency for the hot region.

TABLE 2.1 MIL-HDBK-310. Daily cycle of temperatures and other elements associated with the hottest one-percent temperature values [11]

Time	Temperature		R.H.	Wind (at 3m)		Solar Radiation	
(LST)	(°C)	(°F)	(%)	(m/s)	(ft/s)	(W/m²)	(Btu/h)
1	35	95	6	3	9	0	0
2	34	94	7	3	9	0	0
3	34	93	7	3	9	0	0
4	33	92	8	3	9	0	0
5	33	91	8	3	9	0	0
6	32	90	8	3	9	55	18
7	33	91	8	3	9	270	85
8	35	95	6	3	9	505	160
9	38	101	6	3	9	730	231
10	41	106	5	4	14	915	291
11	43	110	4	4	14	1040	330
12	44	112	4	4	14	1120	355
13	47	116	3	4	14	1120	355
14	48	118	3	4	14	1040	330
15	48	119	3	4	14	915	291
16	49	120	3	4	14	730	231
17	48	119	3	4	14	505	160
18	48	118	3	4	14	270	85
19	46	114	3	4	14	55	18
20	42	108	4	4	14	0	0
21	41	105	5	4	14	0	0
22	39	102	6	4	14	0	0
23	38	100	6	4	14	0	0
24	37	98	6	3	9	0	0

© Department of Defense

CHAPTER 2

Temperature tables used for a hot-day design include solar loading and wind speed to support solar-heated structure temperature and the corresponding passenger cabin heat load calculations. Airplane structure temperatures can far exceed the ambient air in this environment, due to solar heating. Convective cooling (which is proportional to the wind speed), however, limits the structure temperature increase. Since an airport building can prevent wind from striking an aircraft, Table 2.1 wind speeds are irrelevant for commercial aircraft loading passengers at the gate where still air can occur.

Table 2.2 shows the maximum ambient temperatures for the 1-, 5-, and 10-percent frequency of occurrence (during the hottest month), and long-term extreme values corresponding to the maximum temperature predicted to occur at least once during 10, 30, or 60 years.

TABLE 2.2 MIL-HDBK-310. Hot region maximum air temperatures during diurnal [12]

	Highest Recorded	Frequency (hottest month)			Long-term extremes		
		1%	5%	10%	10 years	30 years	60 years
(°F)	136	120	115	113	128	130	131
(°C)	58	49	46	45	53	54	55

© Department of Defense

Table 2.3 shows comparable minimum temperatures for the *extreme cold* region.

TABLE 2.3 MIL-HDBK-310. Extreme cold region minimum air temperatures during diurnal [13]

	Lowest Recorded	Frequency (coldest month)				Long-term extremes		
		1%	5%	10%	20%	10 years	30 years	60 years
(°F)	−90	−78	−70	−65	−60	−86	−89	−92
(°C)	−68	−61	−57	−54	−51	−65	−67	−69

© Department of Defense

The extreme cold temperatures accompany virtually no solar heating and a minimum diurnal temperature swing. The 20-percent frequency temperature data for the extreme cold is included due to the extremely low percentage of aircraft operations occurring at the 10-percent frequency temperatures. Providing a 20-percent frequency temperature on the hot side makes less sense since the OAT regularly exceeds 113°F (45°C) (the 10-percent frequency hot-day temperature) in multiple major airports. Figure 2.4 shows temperature versus altitude data for a range of frequency of occurrences.

The volume of available OAT data can make choosing the appropriate design temperature for each airplane system or structure a confusing task for even veterans of multiple airplane programs. *MIL-HDBK-310* recommends initially considering a 1-percent frequency of occurrence for all climatic elements, except cold temperatures where it recommends a 20-percent frequency. It does not say, however, which 1 percent to use. Is it a regional hot or basic hot, with or without high humidity? This same question pertains to the 20-percent frequency for the lower-end temperatures, with extreme cold, cold, and basic cold with various humidity levels to choose from.

There is no simple formula to follow when trying to build the most cost-effective airplane possible. Instead, the impact of each design temperature change must be assessed based on the total cost to the airplane for expanding operational capabilities, versus the cost to airlines of canceled flights, added ground maintenance procedures, and reduced sales for the aircraft manufacturer.

FIGURE 2.4 High and low atmospheric air temperatures versus altitude (Data from [14]).

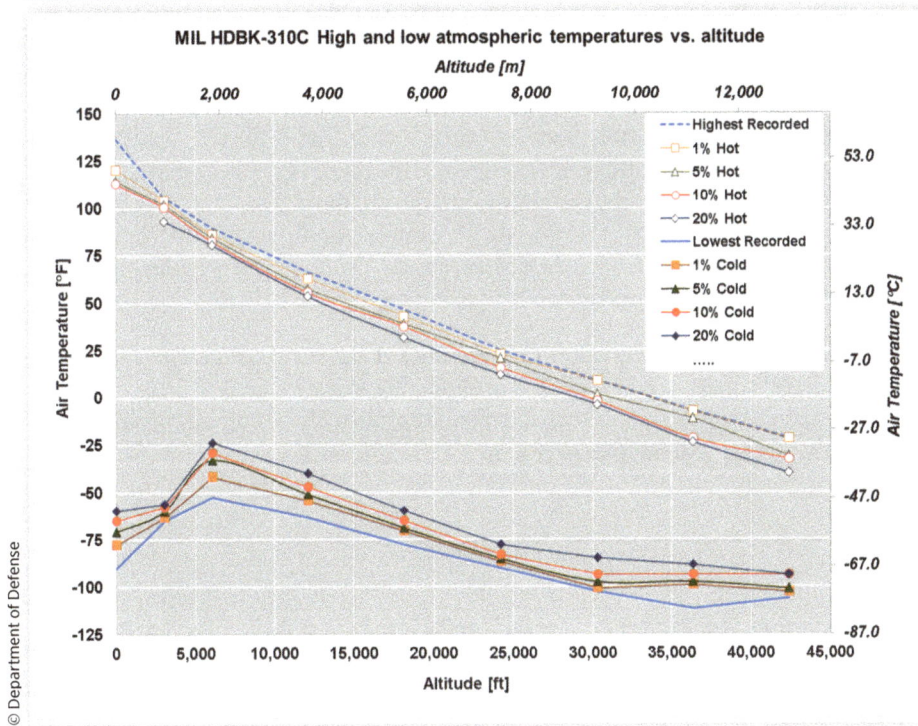

MIL HDBK-310C High and low atmospheric temperatures vs. altitude

© Department of Defense

2.2.2 Boundary Temperature Calculations

After the design OAT selection, the corresponding ground surface and sky radiation and in-flight ram-air recovery temperatures must be determined to support thermal modeling.

Ground and sky temperatures are calculated based on the OAT and corresponding solar loads and *RH*.

2.2.2.1 GROUND

Ground surface temperatures can deviate significantly from OAT for hot-day ambient conditions due to the ground thermal capacitance and solar heating. Both analytical models of the ground and equations can provide estimated surface temperature responses to ambient conditions on a daily and yearly basis.

Analytical Models. Some distance below the surface of the earth, commonly placed at 30 ft (9.1 m), the ground soil temperature is relatively constant year-round, and equal to the water table temperature [15]. Above that depth, the soil/concrete temperature gradually changes at a rate dependent on the thermal characteristics of the material and surface solar heating and convective cooling. The surface color, which determines solar energy absorption, has an especially large impact on the surface boundary, with black asphalt generating the highest temperatures.

A thermal model simulating conductive heat transfer from the constant temperature depth to the surface can predict ground surface temperatures for hot-day simulations.

Since the ground offers a significant thermal lag, the model must simulate the transient response to variations in solar heating, OAT, wind speed, and sky temperature. Ideally, three months or more prior to the summer-time design date (summer solstice), where the solar loading reaches the maximum value (Section 4.2.6.2). The model must also assume a wind speed, which determines the convective heat transfer with the ground surface.

Al-Temeemi and Harris [16] provide subsurface temperature data for Kuwait, a good surrogate for an extreme hot-day environment. This makes the data especially useful for assessing the validity of ground temperature predictions.

Yavuzturk and Ksaibati [17] provide an example of an asphalt thermal model including soil thermal properties, while Marceau and Vangeem [18] include *solar reflectivity* (ρ) data for different concretes.

Equations. The simplest Equation (2.1) assumes the ground temperature follows a *sinusoidal* (sine) function that oscillates around an average value [19]. This same sine function characterizes both the diurnal temperature swing for a single day or the variation in daily average temperature over a year.

$$T(0,t) = T_{ave} + A_0 \sin \omega (t - t_0) \tag{2.1}$$

where

$T(0,t)$ = soil temperature at a depth $z=0$ (the surface) and time t, °F (°C)

T_{ave} = daily or yearly average temperature of the surface, °F (°C)

A_0 = daily or yearly temperature fluctuation amplitude (the range from maximum or minimum to average temperature), °F (°C)

t_0 = time lag from an arbitrary starting point for the minimum temperature, *hrs* for diurnal and *days* for yearly temperature predictions.

The time lag aligns the sine curve maximum and minimum values with the daily diurnal or yearly average ground temperature response.

Diurnal Response. Minimum daily ground surface temperatures occur shortly after sunrise, around 5 or 6 AM. This requires nearly a 180° offset, or 12-hour shift, which corresponds to $t_0 = 12$ (hours), or an hour less for 5 AM.

Yearly Response. Minimum average ground surface temperatures also lag from the minimum OAT, like the daily values. For a minimum average ground temperature occurring on January 30th (in the Northern Hemisphere), $t_0 = 120$ (days). This corresponds to a maximum ground surface temperature occurring six months later, on July 30th.

t = time from the minimum temperature, hours for diurnal and days for yearly temperatures

ω = radial frequency, $2\pi/24\,h$ for diurnal or $2\pi/360$ days for yearly temperature swings The 2π represents a sine wave.

A modified Equation (2.1) captures soil temperature fluctuations below the ground surface.

$$T(z,t) = T_{ave} + A_0 \varepsilon^{-z/d} \sin \left[\omega (t - t_0) - \frac{z}{d} \right] \tag{2.2}$$

where

$T(z,t)$ = soil temperature at a depth z and time t, °F (°C)

z = soil depth, ft (m)

d = *damping depth*, which is the distance below the surface where soil temperature amplitude decreases to $1/e$ (1/2.78) of the soil surface A_0

$$d = \left(\frac{2\alpha}{\omega} \right)^{1/2} \tag{2.3}$$

where

α = thermal diffusivity of the soil, ft^2/h(m^2/h) for the daily diurnal temperature and ft^2/days (m^2/days) for the yearly values.

Equations (2.1) and (2.2) do a reasonable job of estimating the daily and yearly soil temperature variation while introducing errors that Elias et al. [20] address with correction factors.

For cold-day conditions, assuming the ground temperature equals the outside ambient air is usually acceptable. Radiative heat transfer between the ground and aircraft has a negligible effect on airplane structure temperatures for this weather scenario.

Measured Data. Extensive soil temperature measurements are available for agricultural lands in Arizona. This data, however, may not provide the best estimate for extreme hot conditions due to cooler ambient temperatures and differences in ground thermal properties. Dirt has significantly different thermal properties compared to concrete or asphalt designed to hold a 550,000 lbm (250,000 kg) airplane. Therefore, an analytical approach to estimate ground temperatures is normally preferred.

2.2.2.2 SKY

Multiple *sky temperature* (T_{sky}) correlations are available [21] with varying abilities to capture the effects of atmospheric conditions.

The simplest correlations, such as the Swinbank Equation (2.4), are based solely on the OAT [22].

$$T_{sky} = 0.0552 \cdot T_{amb}^{1.5} \tag{2.4}$$

where

T_{sky} = sky temperature, K

T_{amb} = ambient air temperature, K

The Swinbank equation assumes perfectly clear skies with no water vapor or atmospheric gases to affect radiative heat transfer. More complex correlations, which add water vapor and time of day, such as the Berdahl and Martin equation (2.5) [23], are better for correlating analytical models to ground test data.

$$T_{sky} = T_{amb} \left\{ 0.711 + 0.0056 \cdot T_{dp} + 0.0000731 \cdot T_{dp}^{2} + 0.013 \cdot \cos\left(15t\right) \right\}^{1/4} \tag{2.5}$$

where

T_{dp} = dew point temperature, °C

t = hours from midnight, number

2.2.2.3 RAM AIR

The outside boundary air temperature and pressure of airplane structure seen during flight is greater than the OAT and static pressure as the kinetic energy of airplane motion changes to heat, and the airplane compresses the outside air. This occurs in a thin layer of air flowing over the aircraft surface called the *boundary layer*.

Boundary Layer. When an airplane moves, the adjacent air molecules in the airstream are also moving relative to the vehicle surfaces. Fluid flow next to a *bounding* surface, like the wing of a moving airplane, slows due to viscous forces in the fluid causing a velocity variation from zero at the surface (where the flow sticks to structure) to the free stream value (airplane speed) (Figure 2.5).

FIGURE 2.5   Boundary layer visualization showing laminar to turbulent transition [24].

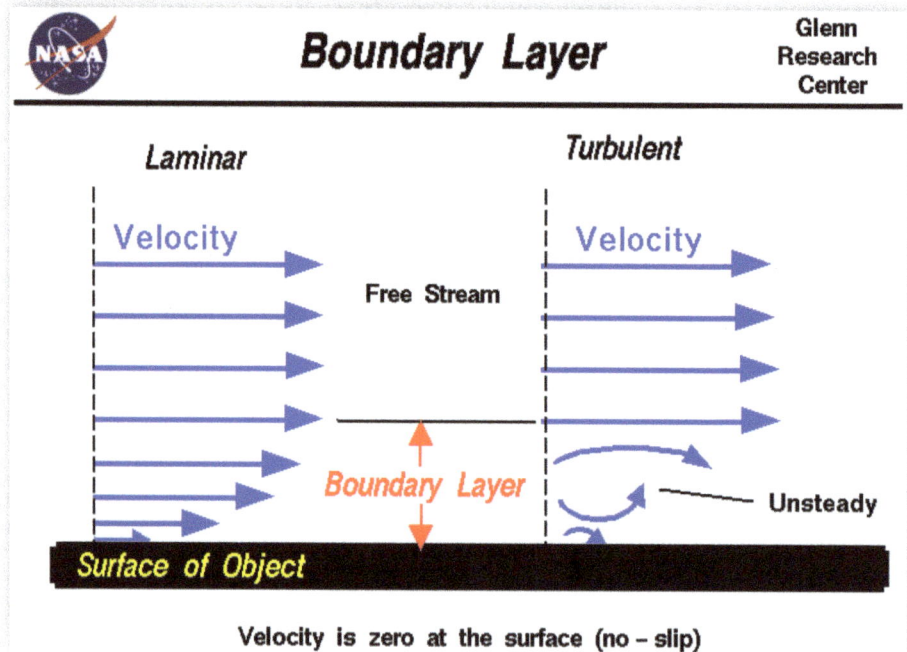

© NASA

Laminar and *turbulent* refer to the orderliness of the streamlines. Laminar flow runs parallel to the surface in smooth streamlines while turbulent air is more erratic, which matters when calculating convective heat transfer or flow resistance along a surface.

Ram-Air Temperature Rise. At the forward-most location where the boundary layer starts, called the stagnation point, all kinetic energy associated with the moving aircraft changes to heat. This generates a warmer (compared to OAT) *total air temperature* (TAT). Any point on the aircraft penetrating the free stream air, like the furthest forward point on a wing leading edge surface or pitot probe mounted off the airplane skin, sees a TAT.

Immediately downstream of the stagnation point, some of the heat generated at the aircraft surface transfers to boundary layer air, leaving less to warm the air contacting the surface. This produces a slightly lower boundary air temperature called the *adiabatic wall temperature* (T_{aw}) [25].

$$T_{aw} = T_s \left(1 + r\frac{\gamma - 1}{2} Ma^2\right)^{\frac{\gamma}{\gamma - 1}} \tag{2.6}$$

where

Ma = Mach number represents aircraft velocity, dimensionless (Section 2.2.3.2 Equation (2.10))

T_{aw} = adiabatic wall temperature, °R (K)

T_s = *static air temperature* (SAT), °R (K)

r = a recovery factor

Laminar boundary layer $0.6 < Pr < 15$ $r = Pr^{1/2}$

Turbulent boundary layer $r = Pr^{1/3}$

where

Pr = Prandtl number, a dimensional property representing the ratio of momentum diffusivity to thermal diffusivity (Section 9.2.1.2). Pr = 0.7 for air.

Laminar and turbulent are determined from a dimensionless parameter (the Reynolds number) calculated from the airplane speed, air properties, and aircraft surface location (Section 9.2.1.2).

The temperature increase above static is the ram-air temperature rise. For ATM, T_{aw} is more useful than TAT, which immediately gives way to T_{aw} upon moving aft from the stagnation point. The T_{aw} equation calculates TAT when the *recovery factor* (r) is one, which defines an isentropic process. No heat transfer occurs to the external environment in an isentropic process.

The aircraft's external skin temperature is calculated from the T_{aw} using a forced convection heat transfer coefficient (Section 9.2.1.2). While external surfaces thermally isolated from aircraft heating (such as an insulated fuselage skin) will approach T_{aw} during cruise, the temperatures can differ significantly during takeoff and climb (two critical thermal design mission times).

2.2.3 Boundary Pressure Calculations

The ram-air temperature rise follows a pressure rise, as the kinetic energy of aircraft movement changes to a dynamic pressure.

$$p_{dynamic} = \frac{\rho v^2}{2g_c} \quad \text{English Units} \tag{2.7}$$

$$p_{dynamic} = \frac{\rho v^2}{2} \quad \text{SI Units} \tag{2.8}$$

$p_{dynamic}$ = dynamic pressure, lbf/ft^3 (N/m^2)

ρ = air density, lbm/ft^3 (kg/m^3)

v = airplane speed, ft/s (m/s)

g_c = gravitational constant, 32.174 lbm-ft/lbf-s^2

2.2.3.1 POUNDS MASS VERSUS POUNDS FORCE

One of the most confusing aspects of the English system of units is that *mass* and *force* are both measured using pounds, even though they have different meanings. Mass is the amount of matter in an object, while force refers to the pressure from the mass due to gravity acting on it. The terms *pound-mass* (lbm) and *pound-force* (lbf) differentiate between mass and force in some technical writings. Other sources convert the density of the fluid used in fluid flow equations to the unit "slugs" by dividing density in lbm/ft³ by a *gravitational constant* (g_c). This latter approach captures the force of gravity acting on the mass by hiding g_c in the density term.

This text uses lbm and lbf rather than slugs since it makes it easier to track units, an effective means of checking calculations for errors.

2.2.3.2 TOTAL (RAM) PRESSURE

The dynamic pressure rise plus the static ambient pressure (Section 2.3, Figure 2.1) produces a *total (ram) pressure* (p_t). For external flow, p_t is usually calculated from the airplane *Mach number* (*Ma*) rather than velocity [26].

$$p_t = p_s \left(1 + \frac{\gamma - 1}{2} Ma^2 \right)^{\gamma/\gamma - 1} \tag{2.9}$$

where

p_t = total pressure, lbf/ft² (Pa)
p_s = static pressure, lbf/ft² (Pa)
Ma = Mach number, dimensionless
γ = ratio of specific heats for a constant pressure and volume process, which equals 1.4 for air

$$Ma = \frac{\upsilon}{c} \tag{2.10}$$

where

c = speed of sound, ft/s (m/s)

$$c = \sqrt{\gamma R T_s g_c} \quad \left(g_c \text{ equals one for SI units} \right) \tag{2.11}$$

where

R = air constant, a factor which relates the physical process a fluid undergoes to a temperature, 53.35 ft-lbf/lbm-°R (287.06 J/kg-K) for air
T_s = static air temperature, °R (K)

Like TAT, the total pressure calculation also assumes a 100% efficient process, with all airplane kinetic energy changing to a pressure rise at the airplane skin or ram-air *heat exchanger* (HX) inlet (Section 5.2.2). The less than 100% efficient conversion process, which generates a lower T_{aw} compared to TAT, has the same effect on the dynamic pressure, generating less pressure than the theoretical value from Equation (2.9).

This pressure difference has a small effect on the airplane's skin external convective heat transfer calculations (Section 9.2.1.2). It can, however, have a large impact on ram-air flow and ventilation rates in unpressurized volumes (pack bay, wheel wells, and leading and trailing edge).

The dynamic pressure rise provides the force to push cooling air through a HX, transferring airplane systems-generated waste heat to the outside ambient air. Accurate pressure recovery factors are, therefore, vital to predicting the aircraft cooling system performance. Efficient inlet and exit designs, which maximize the dynamic pressure recovery, are important to minimizing system weight and size.

The predecessor to the US *National Aeronautics and Space Administration* (NASA) space agency, the *National Advisory Committee for Aeronautics* (NACA), studied inlet and outlet designs to maximize the dynamic pressure, starting in the 1930s. NACA's wind tunnel testing data provided in a 1951 report [27] is still valid today for preliminary sizing. Today, final designs, however, are usually based on *computational fluid dynamics* (CFD) analysis (Section 10.5), a powerful analytical method.

2.2.4 Humidity

Outside ambient humidity degrades AC system performance as heat released in the process of condensing moisture in air raises the AC pack outlet temperature. The heat released (latent heat of condensation/vaporization) is about 970.4 Btu/lbm H_2O (2256 kJ/kg H_2O) [28].

So how big is this effect? An AC pack producing a typical 50°F (10°C) hot-day outlet supply temperature would need to remove more than twice as much energy from the 1-percent hot and humid day, 105°F (41°C) 60% *RH* inlet air, compared to the 1-percent hottest day of 120°F (49°C) and 3% *RH* inlet air.

Higher humidity levels, which occur with lower air temperatures, also decrease the solar load reaching an aircraft structure, reducing its temperature. Aircraft structure, however, must function at the higher temperatures reached under a higher solar load and OAT associated with a dryer atmosphere. *MIL-HDBK-310* includes humidity design data.

2.2.5 Solar Flux

Solar heating drives aircraft skin external temperatures, flight deck ground cooling requirements, and flight crew comfort, and influences passenger cabin temperatures. The maximum value listed in *MIL-HDBK-310*, for a low-humidity environment, is 355 Btu/h-ft^2 (1120 W/m^2) and includes a 24-h diurnal cycle (Table 2.2).

The solar flux data represents a horizontal surface, while aircraft surfaces occur with all orientations. The sun's position generating the *MIL-HDBK-310* horizontal fluxes must, therefore, be determined to calculate the range of solar fluxes received by each external surface for a hot-day design simulation (Section 4.2).

2.2.6 Wind Speed

The speed of wind blowing against an airplane sitting at the airport gate affects the convective cooling rate and subsequent external structure temperature. This effect is particularly strong on the thin fuselage skin, which is thermally isolated from the pressurized volume air by insulation. Higher wind speeds drive structure temperatures closer to the outside ambient, while lower wind speeds or still air allow skin temperatures to rise during solar heating. At night, when clear skies provide a colder radiative heat transfer boundary temperature, lower wind speeds contribute to colder structure temperatures.

2.3 Pressurized Volume

Commercial airplane ambient temperature requirements in the pressurized volume vary between the zones occupied by people, electronics, and cargo. Aircraft operating scenario, OAT, and location in the flight profile also affect the temperature requirements.

Internal ambient air temperatures in the pressurized volume have a much narrower temperature band due to the *environmental control system* (ECS) (Chapter 7). An ECS provides the cooling and heating required to meet the temperature requirement for each zone.

2.3.1 Passenger Cabin and Flight Deck (Thermal Comfort)

Human comfort drives a much more stringent temperature requirement in the passenger cabin and flight deck, compared to other locations on an airplane. An ECS controlling the cabin environment must be able to maintain a temperature near 75°F (24°C) most of the time, with every seat occupied and all systems operational. Slight cabin temperature requirement deviations may be allowed for ambient or operating conditions that rarely occur to avoid oversizing the ECS. Airplane manufacturers also levy additional cabin requirements to ensure passenger and flight crew *thermal comfort*.

The *American Society of Heating, Refrigerating, and Air-conditioning Engineers* (ASHRAE) defines *thermal comfort* as "the state of mind in humans that expresses satisfaction with the surrounding environment." This occurs when the heat generated by the body balances the heat loss, causing the body's skin and core temperatures to provide a sense of neutrality.

Researchers asked human subjects placed in chambers to rate their comfort level, using a *seven-point thermal sensation* scale, while varying their *clothing*, physical *activities*, and thermal *environments*. The average thermal response for many subjects, using the scale in Table 2.4, is converted to a *predicted mean vote* (PMV) equation.

TABLE 2.4 Seven-point thermal sensation scale [29]

Value	Sensation
+3	Hot
+2	Warm
+1	Slightly warm
0	Neutral
−1	Slightly cool
−2	Cool
−3	Cold

Danish professor Povl Ole Fanger developed an energy balance equation to predict the PMV thermal comfort based on the parameters defined in the human testing. The PMV, Equation (2.22), uses a series of calculations defining each thermal comfort parameter, as described in Section 2.3.1.1.

2.3.1.1 THERMAL COMFORT PARAMETERS

Activities. A *metabolic equivalent of task* (MET) unit quantifies the rate of energy release by the body (during an oxidation process), called the *metabolic heat generation rate* (M) [30]. The resting metabolic rate occurring during quiet sitting defines one MET (equal to 58.2 W/m²).

Aircraft thermal comfort calculations often use MET rates developed for buildings, either directly or following minor modifications (Table 2.5). Choosing an equivalent building-level activity for crew and passengers during flight on an aircraft involves some engineering judgment. While *seated relaxed* (1 MET) is the obvious choice for seated passengers at rest, the crew activity level can vary dramatically during a flight corresponding to different categories. Sitting limits a pilot or co-pilot's activity level to a bit more than *seated relaxed*. Challenging flight conditions, such as extreme turbulence, causes internal temperatures to rise during the fight-or-flight bodily response, raising their M. Meanwhile, flight attendants slowly walking through a cabin, serving passengers

TABLE 2.5 Metabolic rates [31]

Activity	MET	W/m²
Reclining	0.8	46
Seated relaxed	1.0	58
Standing relaxed	1.2	70
Standing, light activity (shopping, laboratory, light industry)	1.6	93
Standing, medium activity (domestic work)	2.0	116

drinks, may generate the energy associated with *standing, light activity* (1.6 MET). Their mad dash collecting trash, shortly before landing could, however, increase the heat generation to the *standing, medium activity level* (2 MET).

The flight deck crew can vary the temperature of supply air entering the flight deck and airflow rates from individual supply air nozzles in response to variations in their *M*. Flight attendants, however, moving in the aisles require higher velocity or cooler air contacting their bodies compared to seated passengers located inches away, to offset differences in their *M*. Providing two very different thermal environments so close together is a challenging task for systems designers.

An activity level that generates heat can also generate external *mechanical power*, such as forward movement for a walker. *Work* (W) defines the mechanical power term in the PMV equation. In ATM applications, W is close enough to zero to ignore.

Clothing. The convective and radiative heat transfer from a body strongly depends on the type of clothing and skin coverage. A *clothing unit* (clo) identifies the *clothing insulation* (I_{cl}) thermal effect relative to an easy to relate to standard. One clo defines the insulation required for a person at rest to maintain thermal equilibrium in a 70°F (21°C) room ambient temperature with an air velocity corresponding to a typical ventilation rate [32] (Table 2.6).

TABLE 2.6 Typical insulation values for clothing ensembles [33]

Ensemble description	I_{cl} (clo)
Walking shorts, short-sleeved shirt	0.36
Trousers, short-sleeved shirt	0.57
Trousers, long-sleeved shirt	0.61
Same as above, plus suit jacket	0.96
Same as above, plus vest and T-shirt	0.96
Trousers, long-sleeved shirt, long-sleeved sweater, T-shirt	1.01
Same as above, plus suit jacket and long underwear bottoms	1.30
Sweat pants, sweat shirt	0.74
Long-sleeved pajama top, long pajama trousers, short 3/4 sleeved robe, slippers (no socks)	0.96
Knee-length skirt, short-sleeved shirt, panty hose, sandals	0.54
Knee-length skirt, long-sleeved shirt, full slip, panty hose	0.67
Knee-length skirt, long-sleeved shirt, half slip, panty hose, long-sleeved sweater	1.10
Knee-length skirt, long-sleeved shirt, half slip, panty hose, suit jacket	1.04
Ankle-length skirt, long-sleeved shirt, suit jacket, panty hose	1.10
Long-sleeved coveralls, T-shirt	0.72
Overalls, long-sleeved shirt, T-shirt	0.89
Insulated coveralls, long-sleeved thermal underwear, long underwear bottoms	1.37

The PMV thermal parameter equations use traditional thermal resistance units (R_{cl}), with one clo equal to 0.155 m²-C°/W.

The clothing thermal resistance (R_{cl}) determines the clothed skin ratio [34].

$$f_{cl} = 1.00 + 1.290 R_{cl} \quad \text{for} \quad R_{cl} \leq 0.078 \left(m^2 - {}^\circ C / W \right) \tag{2.12}$$

$$f_{cl} = 1.05 + 0.645 R_{cl} \quad \text{for} \quad R_{cl} > 0.078 \left(m^2 - {}^\circ C / W \right) \tag{2.13}$$

where
I_{cl} = clothing thermal insulation, clo
R_{cl} = 0.155 I_{cl} = clothing thermal resistance (m²-°C/W)
f_{cl} = ratio of clothed/nude surface area, fraction

Environment. Radiative heat transfer between a body and facing surfaces uses a single *mean radiant temperature* (MRT), calculated from the surrounding surface temperatures. MRTs for a sitting or standing occupant are provided in Equations (2.14) and (2.15), respectively [35].

Sitting corresponds to a seated passenger or flight crew member

$$T_r = \frac{0.18 \begin{bmatrix} T_{pr}(up) + \\ T_{pr}(down) \end{bmatrix} + 0.22 \begin{bmatrix} T_{pr}(right) + \\ T_{pr}(left) \end{bmatrix} + 0.30 \begin{bmatrix} T_{pr}(front) + \\ T_{pr}(back) \end{bmatrix}}{2(.18 + .22 + .20)} \tag{2.14}$$

Standing represents a flight attendant serving passengers or working in the galleys.

$$T_r = \frac{0.08 \begin{bmatrix} T_{pr}(up) + \\ T_{pr}(down) \end{bmatrix} + 0.23 \begin{bmatrix} T_{pr}(right) + \\ T_{pr}(left) \end{bmatrix} + 0.35 \begin{bmatrix} T_{pr}(front) + \\ T_{pr}(back) \end{bmatrix}}{2(.08 + .23 + .35)} \tag{2.15}$$

where
T_r = mean radiant temperature (surface temperature body radiates to), °C
T_{pr} = radiant boundary temperatures for surfaces with a parallel view to a body, °C

The v of the air in contact with the body and *clothing temperature* (T_{cl}) determine the convective heat transfer cooling rate. Both T_{cl} and v are solved iteratively following successive guesses of T_{cl} [36].

$$T_{cl} = 35.7 - 0.028 \left(M - W \right) - I_{cl} \left\{ 3.96 \cdot 10^{-8} f_{cl} \left[\left(T_{cl} + 273 \right)^4 - \left(T_r + 273 \right)^4 \right] + f_{cl} h \left(T_{cl} - T_a \right) \right\} \tag{2.16}$$

where
T_{cl} = clothing average temperature, °C
T_a = air temperature, °C
h = convective heat transfer coefficient (W/m²-°C)

The PMV analysis process uses the maximum convective heat transfer coefficient calculated between Equations (2.17) and (2.18) [37].

$$h = 2.38\left(T_{cl} + T_a\right)^{1/4} \tag{2.17}$$

$$h = 12.1\sqrt{v} \tag{2.18}$$

where
v = air velocity in contact with skin (m/s)

RH affects heat transfer from moisture transport due to evaporation and transpiration. A lower RH increases evaporation and transpiration heat transfer, which makes the dry heat of a 100°F (37°C) afternoon ambient air temperature in Phoenix far more comfortable than the similar temperature in humid Singapore.

The moisture transport from a body is proportional to the *partial pressure of water vapor* (p_a) in the air.

$$p_a = p_{sat}\frac{RH}{100\%} \tag{2.19}$$

where
RH = relative humidity, %
p_a = partial vapor pressure, Pa
p_{sat} = saturated vapor pressure, Pa

where:

$$p_{sat} = \varepsilon^{\left[16.6536 - \frac{4030.183}{T_a + 235}\right] \times 1000} \quad [38] \tag{2.20}$$

2.3.1.2 THERMAL LOAD

The difference between the net metabolic heat generation and heat loss to the surrounding environment is the *thermal load* (L) [39].

$$
\begin{aligned}
L = {} & \left(M - W\right) \\
& - 3.05 \cdot 10^{-8}\left[5733 - 6.99\left(M - W\right) - p_a\right] \\
& - 0.42\left[\left(M - W\right) - 58.15\right] \\
& - 1.7 \cdot 10^{-5} M\left(5867 - p_a\right)\left(M - W\right) \\
& - 0.0014 M\left(34 - T_a\right) \\
& - 3.96 \cdot 10^{-8} f_{cl}\left[\left(T_{cl} + 273\right)^4 - \left(T_r + 273\right)^4\right] \\
& - f_{cl} h\left(T_{cl} - T_a\right)
\end{aligned}
\tag{2.21}
$$

where
L = thermal load (W/m²-°C)

The PMV is proportional to the thermal load, with lower values indicating a more thermally comfortable environment [40].

2.3.1.3 **PREDICTED MEAN VOTE**

$$PMV = \left(0.303e^{-0.036M} + 0.28\right)L \qquad (2.22)$$

The PMV equation is valid between −2 and +2 for the following ranges of input parameters.

M 46 to 232 W/m^2 (0.8 to 4 met)
I_{cl} 0 to 2 clo
T_a 10 to 30°C
T_r 10 to 40°C
v 0 to 1 m/s
p_a 0 to 2700 Pa

2.3.1.4 **PREDICTED PERCENTAGE DISSATISFIED**

The PMV predicts the mean value of the thermal votes of a group of people exposed to the same environments with similar clothing and activity levels. A more useful measure for aircraft designers trying to minimize the percentage of passengers who are too cold or too hot is the *predicted percentage dissatisfied* (PPD).

$$PPD = 100 - 95e^{\left(-0.03353\,PMV^4 - 0.2179\,PMV^4\right)} \,[41] \qquad (2.23)$$

where
    PPD = the percentage of people who are thermally dissatisfied, %

The PPD never goes below 5%, even if the PMV equals 0, which indicates that all people would vote feeling thermally satisfied for a condition. This recognizes that it is impossible to please all people with any single set of thermal conditions.

2.3.2 Cargo Compartments

Commercial passenger airplanes contain cargo compartments that carry passenger luggage, cargo, perishables, and live animals. Freighter versions of many commercial passenger airplanes, with cargo containers or pallets replacing seats on the passenger deck, are also manufactured.

Inanimate cargo and live animals have less stringent temperature requirements compared to the passenger cabin. Passenger luggage contains water-based liquids, which freeze at 32°F (0°C), driving the design of sufficient thermal control to prevent freezing temperatures during flight as a minimum standard for most cargo compartments. Some airlines carry live animals and perishables (produce, flowers, etc.) in their cargo compartments that require greater temperature control to arrive at their destination in an acceptable condition. Animals can become overly stressed or die from air temperatures which are too high or too low, while the shelf life of flowers can decrease, providing less time to sell.

Live animal carriage (cargo) needs more restrictive temperature requirements, in addition to *RH* and *carbon dioxide* (CO_2) limits, as shown in Table 2.7 for Boeing commercial airplanes.

TABLE 2.7 Recommended temperature, humidity, and CO_2 requirements for various animal species [42]

Animal*	Desirable temperature range	Recommended relative humidity (RH)	Recommended CO_2
Beef cattle	40–80°F (4.4–26.6°C)		
Dairy cows, mature, dry	40–80°F (4.4–26.6°C)		
Dairy heifers, pregnant	40–75°F (4.4–23.8°C)		
Dairy calves	50–75°F (10–23.8°C)	0–75% RH for swine/ hog	0–0.5% for 1-day-old chicks
Hogs: Over 15 lb	50–75°F (10–23.8°C)		
Hogs: Pregnant gilts	50–70°F (10–21.1°C)	0–80% RH for cattle/ poultry	0–3% for most other animals
Horses	40–80°F (4.4–26.6°C)		
Poultry: Over 10 days old	50–80°F (10–26.6°C)		
Poultry: 1-day-old (unfed)	90–100°F (carton) (32–37°C)		
Sheep	50–75°F (10–23.8°C)		

© Boeing

* Recommended ECS settings are determined based on the type of animals being transported. Source: American Society of Heating, Refrigerating and Air-Conditioning Engineers and Society of Automotive Engineers.

Since the FAA is solely concerned with passenger and crew safety and comfort, trade associations and professional society committees made up of industry experts establish live animal cargo comfort and safety standards. The *International Air Transport Association* (IATA) and SAE International are major contributors, publishing extensive data used to design cargo compartments and establish operational procedures for live animal transport. IATA offers a series of *Live Animal Regulations* (LAR) for "safe, humane transport of animals in air," while SAE publishes *AIR 1600* [43].

2.3.3 Equipment

Electrical and mechanical equipment have a much higher acceptable temperature range than either humans, live animals, or passenger luggage.

Lower lobes contain most commercial airplane electronics in EE bays with active cooling systems (Section 7.4). Their temperature requirements are set by qualification and reliability needs defined in industry standards, such as *MIL-STD-2218 Design Criteria Standard Thermal Design, Analysis and Test Criteria for Airborne Electronic Equipment* [44].

MIL-STD-2218 establishes thermal design, analysis, and verification testing requirements for electrical equipment and cooling systems designed to ensure acceptable equipment performance for all airplane operating scenarios. It states that equipment requiring active cooling must operate within the range of inlet cooling fluid temperatures of Table 2.8, and that outlet fluid temperatures must not exceed 160°F (71°C).

TABLE 2.8 Equipment cooling air and liquid supply temperature requirements [45]

	Air Cooled Equipment				Liquid Cooled Equipment			
	min		max		min		max	
	°F	°C	°F	°C	°F	°C	°F	°C
Storage & non-operating								
	71	−57	203	95				
Conditioned Supply Temperature								
In-flight	−65	−54	131	55	32	0	104	40
ground	−50	−46	131	55	−40	−40	131	55
Bench testing	0	−18	131	55	0	−18	131	55
Ram Air Supply Temperature								
In-flight	−80	−62	131	55				
ground	−50	−46	131	55				
Bench testing	0	−18	131	55				

© Department of Defense

Conditioned supply, for air-cooled equipment, can refer to main deck cabin air exhausted to the lower lobe, where fans use it to ventilate the electronic boxes, or a dedicated cooling system providing chilled air. *Ram-air supply* is outside ambient air supplied by a fan on ground and by the dynamic pressure (Section 2.2.3) available when the airplane is moving in flight. *Liquid-cooled equipment* uses a liquid refrigerant to move heat from the electronics to a heat sink.

MIL-STD-2218 also defines a thermal design process followed on both commercial and military aircraft programs that includes

- Defining the equipment cooling method and coolant flow requirements. Free convection is the preferred method when practical.

- Completing the thermal analysis to ensure adequate cooling for the electronic components for the range of operating conditions.

- Testing to verify component performance and reliability and analytical predictions.

2.4 **Unpressurized Area**

Unpressurized areas on a commercial airplane (Figure 2.6) contain airplane equipment that moves aircraft control surfaces, lowers and raises landing gears and their corresponding doors, stops the airplane (brake), and cools the cabin.

Fuel located in main and center wing tanks, hydraulic fluid lines run from pumps generating high pressures to the equipment, and AC packs located in a fairing underneath the fuselage, also occur outside the pressurized airplane fuselage.

Since unpressurized compartments normally lack dedicated cooling systems, the focus of thermal design in these areas is more toward accurately predicting thermal environments. This supports more realistic (and efficient) design requirements rather than grossly overestimating minimum and maximum ambient temperature exposures.

FIGURE 2.6 Commercial airplane unpressurized volumes.

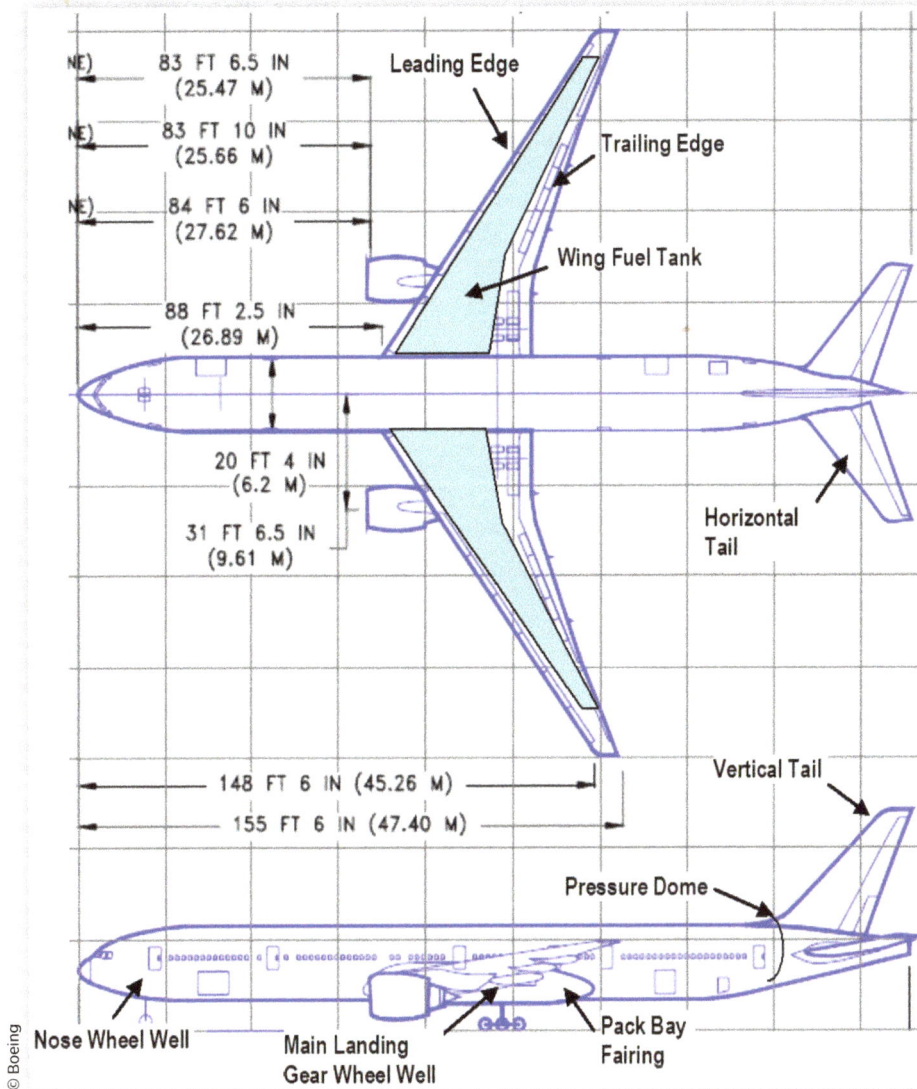

© Boeing

2.4.1 Fuel

The main wing fuel tank's function as a heat sink (Section 5.4.1) for the engine oil cooler and hydraulic system and FAA's fuel flammability requirements set its maximum allowable temperature. High fuel temperatures limit its ability to absorb hydraulic and engine oil cooler heat, potentially degrading the performance of flight critical systems. Higher fuel temperatures also increase the flammability risk. The FAA requires the installation of inerting systems (Section 8.9.2.2) to provide *nitrogen-enriched air* (NEA) to reduce the flammability risk of large commercial airplane fuel tanks. Increasing fuel temperatures can affect the required inerting system size, weight, cost, and power consumption.

Cold fuel temperature limits keep fuel from freezing and starving the engines. Aircraft fluid lines running through a fuel tank can also drive higher minimum allowable fuel temperatures to avoid reductions in system performance. This is primarily a concern for longer high-altitude cruise in extreme cold atmospheric conditions and ground operations at a handful of airports.

Even with cold weather additives, fuel left in the main wing tank can soon go below temperature limits if an airplane spends too much time on the tarmac while fueled in Siberia or Alaska during winter. Operational procedures, such as fueling shortly before takeoff, help minimize disruptions for cold weather operators.

2.4.2 Hydraulics

Fluid and seal capabilities and concerns over heating adjacent structure beyond desired design temperatures limit hydraulic fluid temperatures. *MIL-H-5606H* specifies a hydraulic fluid temperature operating range of −65°F (−54°C) to 275°F (135°C) for military aircraft [46], which normally pertain to commercial airplanes, since they use the same fluids.

Exceedance of the higher temperature is primarily a concern on ground during high ambient and solar loading conditions, following excessive control surface movement during testing.

Hydraulic systems, with lines running through the fuel tanks, are susceptible to reaching very low temperatures from heat transfer to cold fuel during long high-altitude flights over the poles, reducing system responsiveness to commands. While larger diameter tubes can offset the greater resistance to a more viscous fluid, they add weight.

2.5 Structure

The structure must show adequate strength (positive margins), over all aircraft operating conditions.

The minimum load condition, *static*, corresponds to an airplane loaded with fuel and passengers while stationary. *Ultimate* corresponds to the maximum expected load during the airplane life, which can occur anytime after takeoff. *Limit* adds a 1.5 safety factor to the *ultimate* load to account for unexpected conditions, such as severe gusts during a massive storm, and variability in material properties. *Fatigue* refers to the loads associated with repeated load cycling during thousands of flights as an airplane travels from ground to cruise altitude, while *decompression* deals with the rapid loss of pressure following the creation of a hole in the fuselage. Commercial airplane structure must maintain flight following the loss of a window, door, or significant amount of fuselage skin.

Each flight load condition requires corresponding temperature predictions based on an appropriate operating scenario. Structure analysts may pick their initial design temperatures based on past programs. Afterward, a stress analysis will show either positive or negative margins. For negative margins, if acceptable thermal design changes are not possible, the design may require additional material (weight). If positive margins are very high, the structure may be resized to remove weight, leading to an iterative process between the thermal and structural analysts.

2.5.1 External Bulk

Solar heating (Section 4.2), drives external aircraft surface bulk peak temperatures, with darker colored paint generating the highest temperatures. A horizontal black surface facing a peak noon sun in the warmest location defined in *MIL-HDBK-310* can exceed the OAT by more than 100°F (56°C). This assumes calm winds and minimal heat transfer to the back side, which is similar to the insulated crown of a commercial airplane fuselage.

While painting that same surface white reduces the peak temperature close to the OAT, paint limitations that do not accommodate customer livery requests may be unacceptable to an aircraft manufacturer marketing department. With Air New Zealand already flying a black 787 fuselage, paint limitations would place a competitor product under embarrassing scrutiny. Light-colored paints accumulating dirt also absorb more solar energy.

Multiple companies and research centers have been working for decades on dark paints that absorb much less solar energy to address this issue (Section 8.4.1).

Systems-generated heat sources, such as fuselage or wing lights, or heated atmospheric measurement probes (i.e., pitot) can also strongly affect solar-heated structure temperatures.

In-flight, cold-side operating limits are determined by the minimum boundary air temperature, which are close to the T_{aw} (Section 2.2.2.3). The minimum T_{aw} is calculated based on the minimum allowable airplane Ma. Convective heat transfer coefficients are high enough at these airplane speeds to assume structure reaches the T_{aw}.

2.5.2 Internal

Internal structure, which includes floor beams and panels, is thermally isolated from the outside ambient beyond a small section close enough to the airplane's skin for significant conductive heat transfer to occur. The ECS and skin insulation designed to provide controlled temperatures for occupants and equipment, does the same for internal structure. Bulk internal structure (away from airplane heat sources) will, therefore, approach the air temperature over time. Higher structure temperatures also occur next to heat sources, such as power feeder floor beam penetrations or hydraulic tube attachment points.

References

1. Riva, A., "After MH17 Crash, Boeing 777 Safety Record Still among the Best in Aviation History," *International Business Times*, July 24, 2014, accessed January 25, 2017, http://www.ibtimes.com/after-mh17-crash-boeing-777-safety-record-still-among-best-aviation-history-1638354.

2. Weatherbase, "Kuwait City, Kuwait," accessed January 21, 2017, http://www.weatherbase.com/weather/weather.php3?s=28504&cityname=Kuwait-City-Al-Asimah-Kuwait.

3. Weatherbase, "Phoenix, AZ," accessed January 21, 2017, http://www.weatherbase.com/weather/weather.php3?s=87227&cityname=Phoenix%2C+Arizona%2C+United+States+of+America.

4. Weatherbase, "Yakutsk, Russia," accessed January 21, 2015, http://www.weatherbase.com/weather/weather.php3?s=95942&cityname=Yakutsk-Sakha-Russia&units=metric.

5. International Organization for Standardization, "Standard Atmosphere," ISO 2533, 1975.

6. Department of Defense, *Global Climatic Data for Developing Military Products*, MIL-HDBK-310 (Washington, DC: Department of Defense, June 23, 1997), 36.

7. US Department of Transportation/Federal Aviation Administration (USDOT/FAA), Chapter 16, Cabin Environmental Control Systems, *Aviation Maintenance Technician Handbook—Airframe* (2012), 16, 23, Volume 2, Figures 16-39 and 16-40.

8. Williams, M., "What Is the Earth's Average Temperature?" Universe Today, Last modified July 26, 2016, accessed January 12, 2017, http://www.universetoday.com/48328/earth-surface-temperature/.

9. National Research Council, Chapter 2 Environmental Control, *The Airliner Cabin Environment and the Health of Passengers and Crew* (Washington, DC: The National Academies Press, 2002), 35, https://www.nap.edu/read/10238/chapter/4#70.

10. Department of Defense, 1997, 101.

11. Ibid., 72.

12. Ibid., 35, 89.

13. Ibid., 36, 90.

14. Ibid., 35, 36, 89, 90.

15. Reysa, G., "Ground Temperatures as a Function of Location, Season, and Depth," *Build it Solar*, 2015, accessed January 23, 2017, http://www.builditsolar.com/Projects/Cooling/EarthTemperatures.htm.

16. Al-Temeemi, A.A. and Harris, D.J., "The Generation of Subsurface Temperature Profiles for Kuwait," *Energy and Buildings* 33 no. 8 (October 2001): 837–841, Department of Building Engineering and Surveying, Heriot-Watt University, Edinburgh, published in UK.

17. Yavuzturk, C. and Ksaibati, K., "Assessment of Temperature Fluctuations in Asphalt Pavements due to Thermal Environmental Conditions Using a Two-Dimensional Transient Finite Difference Approach," *Journal of Materials in Civil Engineering* 17, no. 4 (August 2005).

18. Marceau, M.L. and Vangeem, M.G., "Solar Reflectance Values of Concrete," *Concrete International* 30, no. 8 (August 2008).

19. Hillel, D., *Introduction to Soil Physics* (San Diego, CA: Academic Press, 1982).

20. Elias, E.A., Cichota, R., Torriani, H.H., and de Jong van Lier, Q., "Analytical Soil-Temperature Model: Correction for Temporal Variation of Daily Amplitude," *Soil Science Society of America Journal* 68 (2004): 784–788.

21. Adelard, L., Pignolet-Tardan, F., Mara, T., Lauret, P. et al., "Sky Temperature Modelisation and Applications in Building Simulation," *Renewable Energy* 15 (1989): 418–430.

22. Swinbank, W.C., "Long Wave Radiation from Clear Skies," *Quarterly Journal of the Royal Meteorological Society* (July 1963): 339.

23. Berdahl, P. and Martin, M., "Emissivity of Clear Skies," *Solar Energy* 32, no. 5 (1984): 663–664.

24. NASA, "Boundary Layer," Glenn Research Center, accessed January 26, 2017, https://www.grc.nasa.gov/www/k-12/airplane/boundlay.html.

25. Kreith, F., *The CRC Handbook of Thermal Engineering* (New York: CRC Press, 1999), 3–35.

26. Roberson, J.A. and Crowe, C.C., *Engineering Fluid Mechanics*, 2nd ed. (Boston, MA: Wiley & Sons, 1975), 472.

27. Frank, J.L. and Taylor, R.A., *Comparison of Drag, Pressure Recover, and Surface Pressure of a Scoop-Type Inlet and a NACA Submerged Inlet at Transonic Speeds*, NACA RM A51H20a (Washington, DC: National Advisory Committee for Aeronautics, December 17, 1951).

28. The Engineering Toolbox, "Fluids—Latent Heat of Evaporation," accessed January 27, 2017, http://www.engineeringtoolbox.com/fluids-evaporation-latent-heat-d_147.html.

29. International Organization for Standardization (ISO), "Ergonomics of the Thermal Environment: Analytical Determination and Interpretation of Thermal Comfort Using Calculation of the PMV and PPD Indices and Local Thermal Comfort Criteria," ISO 7730:2005(E), 11-15-2005, Geneva: ISO, 2, 2005.

30. Wikipedia, "Metabolic Equivalent," 2017, accessed June 31, 2017, https://en.wikipedia.org/w/index.php?title=Metabolic_equivalent&oldid=796909269.

31. ISO, 2005, 18.

32. Wikipedia, "Clothing Insulation," 2017, accessed February 24, 2017, https://en.wikipedia.org/w/index.php?title=Clothing_insulation&oldid=827047787.

33. ASHRAE, Thermal Comfort" chapter in *Fundamentals Volume of the ASHRAE Handbook* (Atlanta, GA: ASHRAE, Inc., 2005).

34. ISO, 2005, 3.

35. Innova Air Tech Instruments, "Thermal Comfort," 1997, accessed June 25, 2017, p. 22, http://www.labeee.ufsc.br/antigo/arquivos/publicacoes/Thermal_Booklet.pdf.

36. Ibid, 17.

37. Ibid.

38. ISO, 2005, 23.

39. ISO, 2005, 3.

40. ISO, 2005, 3.

41. ISO, 2005, 4.

42. Le, L., "Safe Transport of Live Animal Cargo," *Boeing AERO QTR_02.12*, 2012, accessed February 2, 2017, www.boeing.com/commercial/aeromagazine/articles/2012_q2/4/.

43. SAE, "Animal Environment in Cargo Compartments," AIR 1600:1997 (R2011) SAE International, 2011.

44. Department of Defense (DOD), "Design Criteria Standard Thermal Design, Analysis and Test Criteria for Airborne Electronic Equipment," MIL-STD-2218, May 20, 1992.

45. Department of Defense, 1992, 10, 11.

46. Department of Defense, "Military Specification Hydraulic Fluid, Petroleum Base; Aircraft, Missile and Ordnance," MIL-H-6506G, September 9, 1994, 1.

Airplane-Generated Heat Sources

"If you can't stand the heat, get out of the kitchen."

—**Harry S. Truman [33rd US President]**

3.1 Introduction

Occupants and systems generate heat inside an airplane fuselage, which must be removed to maintain an acceptable ambient environment. Occupants comprise the flight crew and, for commercial transports, passengers, and possibly live animals. Systems generating heat include hydraulics, flight controls, fuel, lighting, brakes, and environmental control. Note that the *environmental control systems* (ECS) designed to remove waste heat and cool a fuselage, *electrical/electronic* (EE) cooling, and *cabin air-conditioning and temperature control system* (CACTCS), also generate significant heat loads or high temperatures. For example, EE cooling exhaust air can raise cabin air temperatures, increasing the *air-conditioning* (AC) pack load. Heat transfer from AC pack components may also cause adjacent structure, or transport elements (like wire runs), to exceed desired design temperatures.

Some aircraft systems generating heat inside the fuselage also transfer heat to external aircraft structure and equipment in the unpressurized areas, such as wings, the *main landing gear* (MLG) wheel well, and AC pack bay. The details of major aircraft-generated heat source follow.

3.2 Occupants

Passenger, flight crew, and animal metabolic processes generate energy that the body rejects to the surrounding environment as *sensible* and *latent* heat.

3.2.1 Sensible Heat

Sensible heat is the energy required to change the temperature of a substance with no phase change. For passengers sitting in an airplane cabin, it refers to the energy transferred from their warmer skin surfaces through radiation, convection, and conduction heat transfer (Sections 9.2.1) to the cooler air, surrounding structure, and seats, respectively. Sensible heat provides a cabin heat load AC packs must remove.

Sensible heat generation depends on air temperature, activity level, and, to a much lesser extent, *relative humidity (RH)*.

3.2.2 Latent Heat

Latent heat is the energy exchange (with no temperature change) as a material undergoes a phase change. For airplane occupants, it refers to the energy released when the AC packs condense moisture in their supply air source. Condensing moisture in outside ambient air during ground operations can consume a significant portion of the pack cooling capacity. Energy released in changing from a vapor to liquid is the heat of fusion.

3.2.3 Passengers and Crew

We often base passenger and crew metabolic heat loads used in aircraft design on building occupant estimates from the *American Society of Heating, Refrigerating, and Air Conditioning Engineers* (ASHRAE) (Table 3.1).

Thermal comfort calculations share similar activity categories (Table 2.5, Section 2.3.1.1) with Table 3.1. Only the sensible load acts as an aircraft cabin heat load in heat transfer calculations and computer models. The latent load, however, determines the cabin *RH*, which can affect AC pack cooling performance (Section 7.2.2) and thermal comfort (Section 2.3.1).

TABLE 3.1 Average metabolic rate – male adult, 75°F (24°C) dry bulb temperature [1]

Degree of activity	Sensible Btu/h	(W)	Latent Btu/h	(W)
Seated at rest	228	(67)	113	(33)
Seated, very light work	239	(70)	171	(50)
Standing, walking slowly	239	(70)	205	(60)
Moderate work	273	(80)	273	(80)
Heavy work	525	(154)	941	(276)

© ASHRAE

3.2.4 Live Animal Cargo

Carrying live animals in an airplane cargo compartment can be very profitable for an airline. The number of animals a cargo compartment can safely carry depends on the cargo compartment design and the animal heat generation rate (Figure 3.1).

Sealed cargo compartments to minimize the leakage of a fire suppressant, which suppresses flames and reduces the compartment oxygen content during a fire event

FIGURE 3.1 Sensible heat production rates [2].

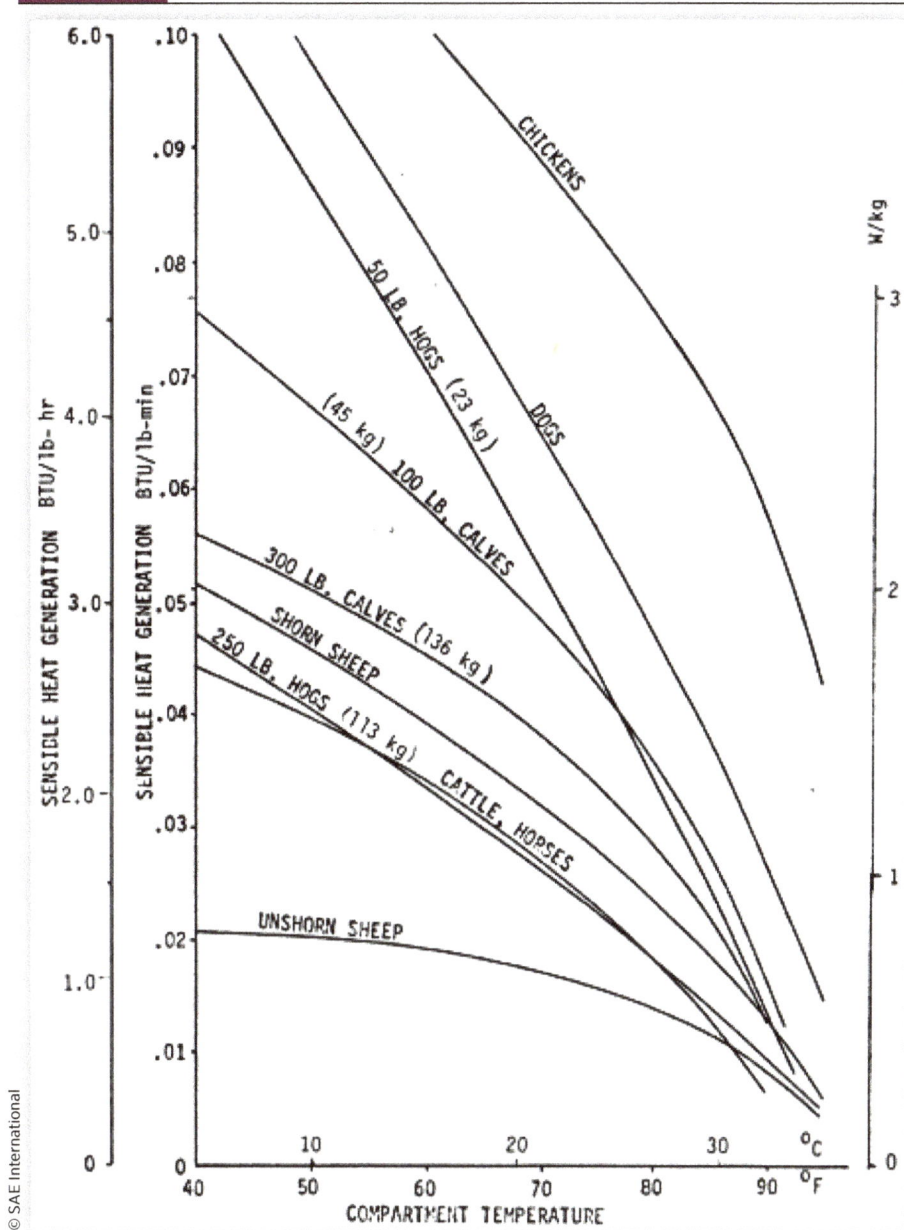

(Section 6.2.3). Therefore, cargo compartments lacking dedicated ventilation can only carry a few animals before exceeding acceptable *carbon dioxide* (CO_2) levels or maximum air temperatures.

Some cargo compartments have dedicated AC systems (Section 7.2.4) to provide conditioned air, significantly increasing live animal cargo carrying capacity. Cargo compartment AC systems using temperature-controlled pack air or a separate refrigeration system add another airplane heat load to account for in the ATM design.

The live animal latent heat load can have a much larger effect on animal comfort compared to passenger latent head loads because of the much denser packing. While

© SAE International

some no-frill carriers would love duplicating such seating arrangements in their cabins, passenger evacuation requirements prevent them. Commercial airplanes must demonstrate the ability to evacuate all passengers and crew within 90 s, to meet US *Federal Aviation Administration* (FAA) certification requirement *AC 25.803-1A*.

No such requirement exists for animals in a cargo compartment. At the higher animal densities, like a cargo compartment filled with hogs, the *RH* can get too high to support sufficient evaporative cooling for what otherwise would be an acceptable temperature. Water changing from a liquid to vapor (sweat evaporating from mammal skin) removes heat, cooling the animal's surface.

Table 3.2 illustrates the effect of *RH* on the equivalent dry bulb temperature for hogs.

TABLE 3.2 Effective compartment temperature increase to allow for high humidity effective temperature increase [3]

Relative humidity	Dry bulb temperature					
	°F (°C) 79 (26.0)	°F (°C) 82 (28.0)	°F (°C) 86 (30.0)	°F (°C) 90 (32.0)	°F (°C) 93 (34.0)	°F (°C) 97 (36.0)
70%	1.3 (0.7)	2.2 (1.2)	3.1 (1.7)	4 (2.2)	5 (2.8)	6.1 (3.4)
80%	2.2 (1.2)	3.6 (20)	5.2 (2.9)	6.7 (3.7)	8.6 (4.8)	10.4 (5.8)
90%	3.6 (2.0)	5.8 (3.2)	8.3 (4.6)	10.8 (6.0)	13.9 (7.7)	17.3 (9.6)
100%	5.9 (3.3)	8.6 (4.8)	11.9 (6.6)	14.9 (8.3)	19.3 (10.7)	28.8 (16.0)

© SAE International

An effective compartment temperature captures the reduced cooling available for a dry bulb (air) temperature. For example, Table 3.2 shows that a cargo compartment air temperature of 86°F (30°C) combined with an *RH* of 90% would provide the same thermal stresses as compartment air temperature 8.3°F (4.6°C) higher with a lower *RH*. In other words, the cargo compartment exhibiting these ambient conditions would be unacceptable for animals requiring a transport temperature less than 94.3°F (34.6°C).

3.2.5 Avionics and Electrical Equipment

Avionics are the electronic systems used on aircraft to help the vehicle get from one place to another in the safest most efficient and comfortable manner possible. They support functions such as communication, navigation, weather reporting, collision avoidance, flight controls, cabin temperature and pressure control, *in-flight entertainment* (IFE), and the infamous "black boxes" collected following a crash. The "black boxes," which are orange for better visibility, contain recorded flight parameters and flight deck conversations for the most recent two hours of flight. These are the key data source for determining the cause of airplane crashes.

Military aircraft trade the IFE functionality for weapons systems and anti-weapon countermeasures.

Removing heat generated from electrical current flowing through electronics prevents premature component failures from high temperatures. Avionics and electrical equipment power dissipation numbers suppliers provide, which do not capture in-service equipment usage, may require modifications based on knowledge gained from the in-service fleet.

The power consumption of most electrical equipment on an aircraft varies widely during a mission. Some equipment, such as microwave ovens used to reheat meals, operate in a small percentage of the mission time. Estimating the predicted equipment usage over each mission segment to identify the peak heat load (sizing point) helps avoid grossly oversizing the cooling system. This is another area requiring significant engineering judgment without comprehensive in-service data.

3.3 Flight Controls and Hydraulic Systems

Aircraft flight control systems include external surfaces (Figure 3.2) that move to provide lift or control an aircraft's direction in flight, linkages to use the power supplied to move the aircraft control surfaces, computers to control the surface movement based on commands from the cockpit or aircraft sensors, and the cockpit controls and displays.

FIGURE 3.2   Airplane control surfaces and function [4].

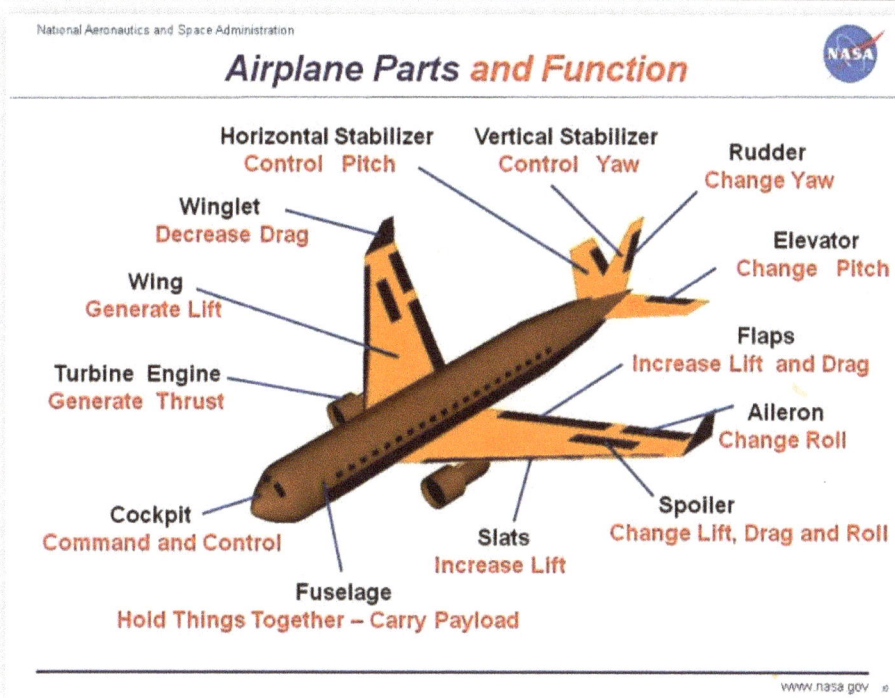

Airplane flight control surfaces include slats and flaps that extend to increase the wing area and provide more lift at low speeds, and primary flight control surfaces to change the airplane direction and ensure stable flight.

Aircraft design must account for the heat generated by systems moving flight control surfaces. The actuator movement takes energy, usually generated by a hydraulic system or engine-driven electrical generators, and transfers it to the control surface actuators using fluid in hydraulic system tubes or electrical current in power feeders. Flight control actuators have redundant power systems for occupant safety, which a combination of electric and hydraulic systems can supply.

3.3.1 **Hydraulic**

Hydraulic systems use a pressurized fluid to power actuators, which move flight control surfaces (Figure 3.2) and operate landing gears and their corresponding doors, pack bay doors, thrust reversers, brakes, and wheel anti-skid systems. The hydraulic systems consist of a reservoir to accumulate fluid, a pump to increase the fluid pressure, which provides energy to move actuators, a *heat exchanger* (HX) to transfer waste heat to a heat sink (usually fuel), and tubing to carry the fluid to and from locations where work occurs.

A hydraulic system generates heat by subjecting a fluid to a very high pressure, typically 3000 to 5000 psi, and from running the fluid through an actuator doing work. Frictional losses from a continuous, but much lower, *leakage* flow also generates waste heat. The higher-pressure design, which provides more force over a smaller cross-sectional area (to reduce the tubing diameter and actuator size), also generates more heat.

Hydraulic tubing often runs through the lower lobe of the fuselage, from the pack bay at the wing to the nose landing gear (on commercial transport airplanes). The tubing transfers heat to the surrounding air and airplane structure through convection and radiation. Heat gained by the air, however, is minor relative to other cabin heat loads because the tubes are so small. The impact on structure temperatures can be high, however.

Flight control operations impact ATM as control surface actuator movement generates large amounts of heat, which can raise the temperature of adjacent structure or systems. Hydraulic actuators transfer most of that heat to the hydraulic fluid supplying the power. Fluid leaving the actuator transfers that waste heat to its surroundings, and a fuel tank HX, on its way to a hydraulic reservoir.

3.3.2 **Electric**

More electric aircraft (MEA) system architectures replace much of the hydraulic functionality with electric motors with electrical wires replacing hydraulic tubes. Electrical wires (and electrically driven actuators) can also generate high temperatures, which can negatively affect adjacent structure. The electric actuators provide a less complex thermal design, compared to hydraulic systems, since they eliminate the need for a fuel tank heat sink (Section 5.4.1) and leakage flow heat transfer calculations.

3.3.3 **Flight Control Thermal Impact**

The frequency of control surface movement, external ambient conditions, and aircraft design requirements determine the thermal impact of flight controls on the aircraft.

Faster actuator movements for a longer time produce the most heat, which transfers to the surroundings through radiation and convection heat transfer. This can occur during prolonged flight through a turbulent atmosphere as the flight control system works to provide a smooth ride. Maintenance or preflight checkouts on ground shortly before takeoff can also generate actuator movements producing sufficient heat to impact the airplane design. The higher potential ground OAT and much lower convective cooling rates can more than offset a longer period of continuous actuator movement in-flight with more ambient cooling, making ground operations the critical design point.

3.4 **Lights**

Commercial aircraft include both external lights located outside the pressurized volume and internal lights in the pressurized volume areas accessed by passengers, flight, or maintenance crew.

FAA required external lighting on commercial airplanes include an anti-collision system of white strobe lights, which project 360°, and colored position lights indicating the left and right side of the airplane (*AC 20-30B*). External door lights also provide illumination during an emergency. Additional lights in the MLG wheel well support maintenance personnel and the preflight pilot walk-around.

Internal lights include the passenger cabin overhead multicolored *light-emitting diode* (LED) lights used to control the ambience, individual lights at each passenger seat and flight crew work area, cargo compartment lights used while loading and unloading cargo, and maintenance personnel lighting in the EE bays.

Cabin and crew station lights (and supporting electronics), including those in the flight deck, provide the only cabin light heat load for AC pack sizing. The remaining lights outside the pressurized volume are assessed for their thermal impact on structure to determine appropriate ambient environments to assume in calculating their reliability.

LED lighting, which covers most lighting applications on large commercial airplanes, is more sensitive to temperature increases compared to incandescent, fluorescent, or halogen bulbs. Determining an accurate lifetime ambient environment for an external light is, therefore, more critical on today's latest aircraft compared to past models. While LEDs generate much less waste heat, compared to previous lighting options, the power density of the electronics driving the lights can be high enough to negatively impact the mounting structure, driving a continuing need for thermal modeling of structural interfaces.

3.5 **Power Feeders**

Power feeders are bundles of wires carrying electricity from the engine generators, *auxiliary power unit* (APU) generators, or ground connection to power panels in the EE bay. Smaller wire bundles and individual wires distribute electrical power from the power panels to equipment throughout the airplane.

An electrical connection to ground source power or portable ground carts or an aircraft-mounted APU generator provides electrical and/or pneumatic power to run airplane systems during passenger loading. Ground power usage is preferred since APUs are less energy efficient and a major source of air pollution at airports. An APU can also provide in-flight power to restart engines or supply flight critical equipment in an emergency.

An ECS removes waste heat generated by electrical current flowing through wires, as calculated in Equation (3.1).

$$P = I^2 R_{elec} \tag{3.1}$$

where
   P = power (waste heat), J-s, or W
   I = current, amps
   R_{elec} = wire electrical resistance, W/amps^2 (ohms)

Thermally efficient designs (that minimize heat generation) would tend to use *heavy* and *expensive*, *low electrical resistance* and larger diameter copper wires. Conversely, a wire configuration optimized for *weight* and *cost* would run *higher (electrical) resistance* aluminum wires that are *lighter* and *cheaper*. Weight and cost are the two most important airplane metrics tracked (unlike waste heat). Thermal designers, therefore, continually face the challenging task of finding design solutions for aircraft configurations that produce the greatest amount of waste heat, in this case, aluminum wiring carrying a high current.

3.5.1 Electromagnetic Interference

The total waste heat generated by multiple closely spaced wires in the power feeder can be higher than the values calculated in Equation (3.1) due to inductive loads from shielding designed to limit *electromagnetic interference* (EMI).

A current running through a wire generates a magnetic field which generates a current in an adjacent wire. Electrical currents generate magnetic fields while magnetic fields generate currents in metals. When that adjacent wire contains a signal (which consists of electrons), the additional electrons generated from that magnetic field can interfere with the sent messages, producing EMI. EMI can generate faulty signals that ruin the picture quality for wires streaming a video to a seatback screen or generate erroneous commands causing flight control surface actuators to move erratically.

A lightning strike can generate huge amounts of EMI, knocking out the entire electrical power system in an aircraft lacking protective measures. To avoid EMI, power feeders may include shielding (a wire mesh), which captures stray magnetic fields while generating an inductive (thermal) load (Section 8.12).

3.5.2 Inductive Loads

The magnetic field generated by the current flowing through one wire, which creates a current and heat load on an adjacent conductor (wire or shielding), generates an inductive (thermal) load. Energy that generates heat in the adjacent wires is no longer available to power the systems supplied by the power feeder. This reduction in power quality is a concern to the electrical group while the heating of the surrounding wires is more important to the thermal team, which must account for it in analytical modeling.

Specialized analytical tools and testing are the main means for predicting inductive losses. Both testing and analysis help identify arrangements of adjacent wires that will cancel some of the magnetic fields responsible for generating inductive power losses.

3.6 Brakes

Airplane brakes stop an airplane during landing or a *rejected takeoff* (RTO) and provide speed control during taxi under engine power. An RTO occurs when the flight crew slams on the brakes to avoid hitting something on the runway or to stop taking off due to an emergency, such as an engine failure. The airplane brakes operate like a car's, converting most of the energy of motion to frictional heat from rotating disks (rotors) contacting brake pads (stators). The rotors and stators alternate to form a stack of contacting disks made of stainless steel or carbon that absorb the heat, which increases their temperature (Figure 3.3).

© Boeing

FIGURE 3.3   Airplane brakes [5].

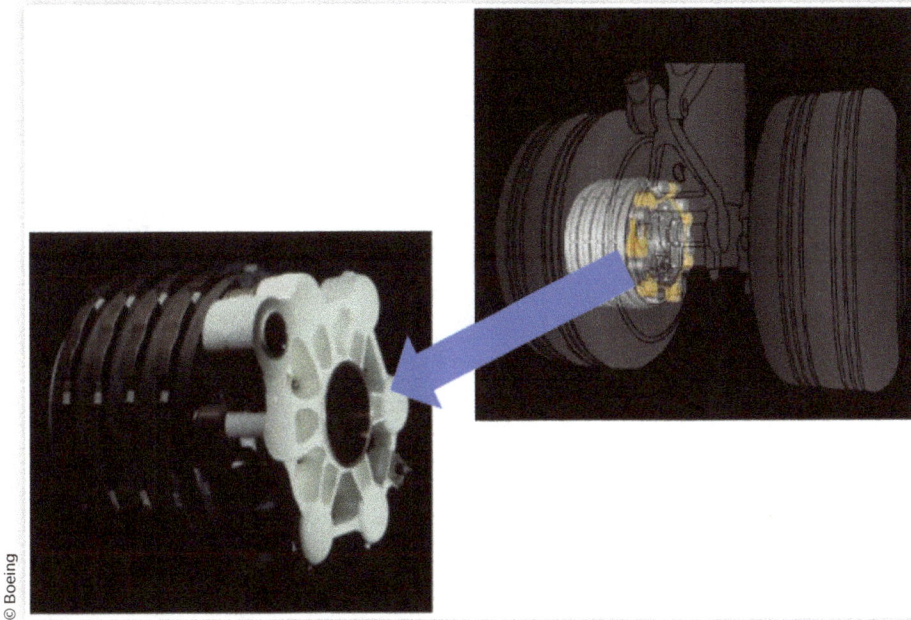

3.6.1 Brake Heat Sink

The *brake heat sink mass* (stator) is based on the maximum *aircraft mass* at the maximum aircraft ground speed prior to initiating braking, the *thermal capacitance* of the brake heat sink material, and maximum allowable *brake temperature*. This relationship comes from a version of the first law of thermodynamics.

$$\frac{1}{2} m_{aircraft} v_g^{\ 2} = \left\{ mc_p dT \right\}_{brake} \tag{3.2}$$

where

$m_{aircraft}$ = maximum aircraft mass (including payload) at the peak speed the aircraft is designed to stop from, lbm (kg)

v_g = aircraft ground speed when brakes are first applied, ft/s (m/s)

c_p = brake material specific heat, Btu/lbm-°F (J/kg-°C)

m = brake heat sink mass, lbm (kg)

dT = brake heat sink temperature rise during braking, °C (°F)

3.6.2 Brake Temperatures

Since weight is so important in aircraft design, the maximum design temperature of commercial airplane brakes can far exceed the reported 392 to 752°F (200 to 400°C) auto brake temperatures for a typical controlled mountain grade descent [6].

The Boeing 777 center stator temperature illustrates the magnitude of the aircraft brake temperatures, by exceeding 1000°C (1832°F) 40 s after landing (Figure 3.4) for a design condition.

Peak stator temperature measurements come from brakes instrumented for lab and airplane flight testing. Production airplane *brake temperature monitor sensors* (BTMS) are placed away from the hot contact surface, to avoid wearing away with the brake pads.

FIGURE 3.4 777 Brake temperature versus time [7].

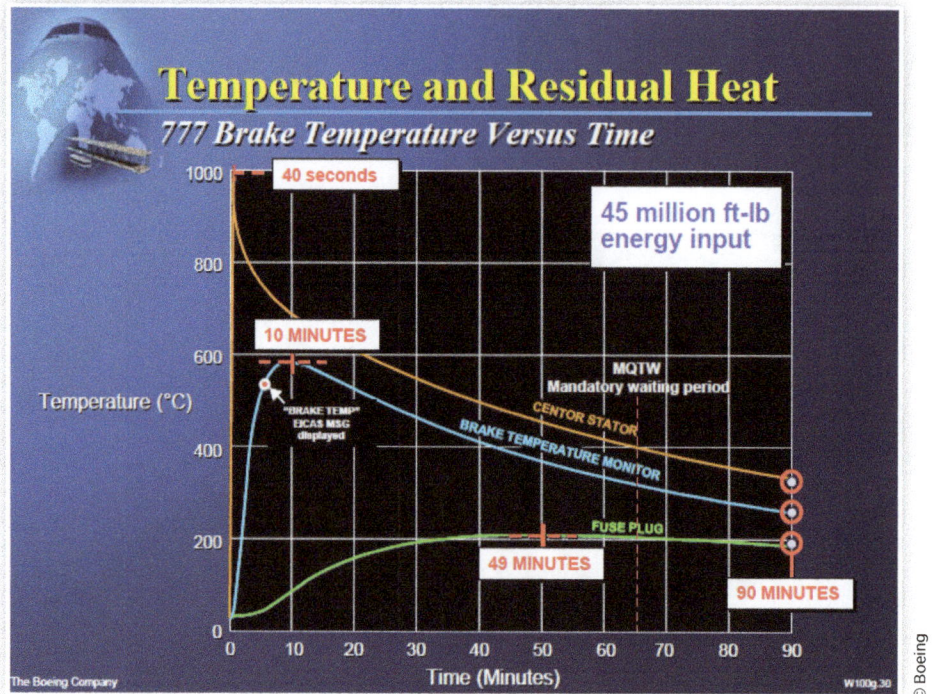

There they measure a lower, and delayed, temperature response to heating due to the thermal capacitance of the stator material. Conductive heat transfer from the contacting brake pads to adjacent brake mass soon lowers peak brake temperatures. The heat is then dissipated to the surrounding ambient air by radiation and convection. While brakes include heat shields to limit heat transfer to wheel well structure and equipment from a retracted landing gear, some radiant energy escapes from gaps in the shield (designed to allow ventilation airflow).

3.6.3 Brake Heating during Successive Missions

Peak stator temperatures, which occur at the end of landing, depend on the amount of braking energy captured relative to cooling occurring during successive missions.

Landing, taxi, and stopping increase the brake (stator) temperatures in capturing the kinetic (braking) energy of a moving aircraft. The energy of braking exits the stator from convective and radiative heat transfer to the outside ambient while the MLG is extended and to the wheel well when retracted.

Stator temperatures can ratchet up between missions with insufficient time with no heat input between landings to cool, due to frequent short flights with little time at the gate. This is a common concern for Boeing 737 and Airbus A320 operators who keep fares low through rapid turnarounds supporting high utilization rates for short-hop route networks. Long taxi runs (especially downhill), with multiple stops or RTOs, can accelerate stator heating, generating the highest brake temperatures. Excessively hot brakes can degrade stopping performance and overheat MLG wheel well structure and equipment.

Brake heat transfer from an extended landing gear of a moving aircraft provides little thermal impact to aircraft structure due to the location away from the aircraft structure and systems, and ample air movement to dissipate the heat. Once the aircraft stops at the gate (following landing and taxi), the plume generated by hot brakes could possibly overheat components in its path.

3.6.4 MLG Wheel Well

After takeoff and gear retraction, the thermal threat to aircraft structure and adjacent equipment is much higher due to limited airflow to remove hot air accumulating in the MLG wheel well. Reductions in structure strength and equipment malfunction or failure due to high temperatures are a primary concern. While ambient ventilation occurs through holes designed to drain liquids and equalize pressures between the outside ambient and MLG wheel well, it may be inadequate to maintain the desired structure and equipment temperatures.

One means of protecting MLG wheel well structure and systems from excess brake heat is to provide adequate time for the brakes to cool between each landing and taxi cycle. Flight procedures may prevent an airplane from leaving the gate until the BTMS temperature falls below a value designed to prevent unsafe heating of the MLG wheel well structure and equipment (after gear retraction).

3.6.5 BTMS Selection at Gate Release

A sufficiently low BTMS temperature at gate release anticipates brake heating occurring during taxi and stopping to avoid excessive brake temperatures at gear retraction (following takeoff). The lowest engine settings can produce too much thrust to maintain a 19–25 mph (30–40 km/h), or less, taxi speed without braking. Airplane engines designed to produce the thrust required for flight force the pilot to ride the brakes during taxi.

The BTMS value at gate release must also anticipate brake wear that reduces the mass available to absorb the heat. While the brake manufacturers are responsible for the calculations required to calibrate BTMS settings, understanding these details is important to the aircraft manufacturer thermal analysts responsible for modeling brake heating of a retracted MLG. Predicting the thermal effects of brake heating on the MLG wheel well structure and equipment is a critical and challenging ATM task.

3.6.6 Brake Fans

Delayed takeoffs, due to hot brakes, have driven the development of cooling fans that blow ambient air over the brakes. The Safran Corporation builds brake cooling fans that install on the wheel axle for most Airbus commercial aircraft [8].

Ground cooling fans are also available to accelerate brake cooling, which airlines reluctantly use due to the cost and inconvenience of an added maintenance procedure.

3.6.7 Thrust Reverses

Most aircraft jet engines also have thrust reversers to deflect engine air forward, reducing the energy absorbed by the brakes during landing and the corresponding temperature rise (Figure 3.5). Thrust reversers also reduce brake wear.

FIGURE 3.5 Thrust reversers [9].

© Pieter van Marion/CC BY-SA 2.0

3.6.8 Fuse Plugs

Aircraft tires include fuse plugs that melt to deflate the tire following overheating. This prevents a tire blowout from excessive thermal expansion of the nitrogen in the tire during taxi and takeoff. Thermal modeling defines operational procedures to prevent this hazardous situation from occurring.

3.7 Environmental Control System

Systems that supply air to pressurize the cabin, provide cooling, or remove heat also produce waste heat, which can raise adjacent structure and equipment temperatures. Some systems can also increase the cabin heat load. Other systems, which use heat to prevent unsafe levels of ice from building up on flight critical equipment or external skin, provide a potential threat to aircraft structure durability.

3.7.1 Air Supply

Compressed outside ambient air pressurizes the aircraft cabin and provides power to support system functions, such as running AC packs or starting engines. The theoretical temperature rise for air passing through a compressor follows [10].

$$T_{t2} = T_{t1} \left(\frac{p_{t2}}{p_{t1}} \right)^{\frac{\gamma-1}{\gamma}}$$

(3.3)

where

T_{t*} = total air temperature, °R (K)
p_{t*} = total air pressure, any consistent unit
$_{1,2}$ = conditions before and after compression
γ = ratio of specific heats, unit less

The p_{t1} and T_{t1}, refer to outside ambient air total pressure (Section 2.2.3.2) and adiabatic wall temperature (Section 2.2.2.3).

The required supply pressure (p_{t2}) can extend beyond the amount needed to overcome the flow resistance in the ducting network to provide power to run AC packs. An *air cycle machine* (ACM) (Section 7.2.2.1) (in an AC pack) uses high-pressure air, as a power source and refrigerant in an open Brayton cycle, to provide temperature-controlled air. Providing higher-pressure air to power an ACM, in addition to pressurizing the cabin, significantly increases the supply temperature (Table 3.3).

TABLE 3.3 Outside ambient temperature rise during pressurization

Alt	35,000 ft	(10,668 m)		
Mach	0.85	0.85		
Pstatic ambient	3.47 psi	(23.9 kPa)		
Ptotal ambient	5.56 psi	(38.3 kPa)		
Tstatic ambient	−26°F	(−32°C)	std+41°F day	
Tadiabatic wall	26°F	(−3°C)		
	Air cycle machine		Vapor cycle machine	
Pbleed	45.0 psi	(310.3 kPa)	14.0 psi	(96.5 kPa)
Pratio	8.1	8.1	2.5	2.5
Tbleed	424°F	(218°C)	173°F	(78°C)

© SAE International

For flight at a 0.85 Mach number at 35,000 ft (10,668 m), increasing the static ambient pressure to 45 psi (310 kPa) increases the ambient bleed temperature to 424°F (218°C). This assumes a warmer ambient environment than typically seen, although not the hottest.

These temperatures may exceed the fuel autoignition temperature, while structure designed to withstand such high temperatures would be extremely heavy. Therefore, a HX (precooler) cools the hot bleed air before it flows down a bleed-air duct located in the leading edge, forward of the wing front spar. The reduced temperature air is still too hot to mount a duct without thermal protection for surrounding structure. High temperatures continue as the duct enters the AC pack bay.

AC cooling systems not using bleed pressure as a power source, like a *vapor cycle machine* (VCM) (Section 7.2.2.2), reduce the needed bleed pressure and corresponding temperature rise. Table 3.3 shows a 14 psi (96.5 kPa) bleed pressure lowering the bleed temperature to 173°F (78°C), eliminating the need for a precooler.

3.7.2 Packs

An AC pack includes ducts carrying bleed air and equipment with hot surfaces, such as compressors, which heat the pack bay air and adjacent *center wing tank* (CWT) surfaces and fairing. Fuel tank temperature increases raise the fuel/ullage temperature, increasing the flammability risk, which is normally mitigated by reducing the oxygen content of the air entering the fuel tank using a *nitrogen-generating system* (NGS) (Section 8.9.2.2).

3.7.3 Fans

The air distribution, recirculation, EE cooling, and *lavatory* (lav)/galley vent system use fans that add large amounts of heat, primarily to the air they move.

3.7.4 Anti-icing/Deicing Systems

Aircraft include systems for preventing or removing ice that can jeopardize flight safety or aerodynamic performance. Critical areas requiring ice protection include flight monitoring instrumentation, such as pitot probes (which determine the aircraft speed), the wing leading edge, and engine inlet (Section 7.5).

Hot bleed air or electric resistance heating raises surface temperatures high enough to either evaporate striking water droplets before they freeze or break the bounds of ice after it has formed on the surface. Structure affected by heating at these critical areas must safely accommodate elevated temperatures.

References

1. Parsons, R., *ASHRAE HANDBOOK: Fundamentals: Inch-Pound* (January 7, 2005), ISBN-10: 1931862702.

2. SAE International, "Animal Environment in Cargo Compartments," SAE International Aerospace Information Report, AIR1600 Rev. A. Reaffirmed 2015-11-01, 2015, 15.

3. Ibid., 13.

4. National Aeronautics and Space Administration (NASA), "Airplane Parts and Functions," 2015, www.nasa.gov. Last Updated: May 05 2015, accessed February 11, 2017, https://www.grc.nasa.gov/www/k-12/airplane/airplane.html.

5. Root, R., "Brake Energy Considerations in Flight Operations," Boeing Commercial Airplanes, September 2003.

6. Glennon, J.C., "Brake Failure Analysis," Crash Forensics, January 5, 2017, accessed February 12, 2017, http://www.crashforensics.com/brakefailure.cfm.

7. Root 2003.

8. Safran, "Our Brake Solutions," Safran Ventilation Systems: Brake Cooling, accessed February 18, 2017, https://www.safran-electrical-power.com/ventilation-systems/our-brake-cooling-solutions.

9. Wikipedia, "Thrust Reversers," 2017, accessed February 19, 2017, https://en.wikipedia.org/w/index.php?title=Thrust_reversal&oldid=782596770.

10. Gas Processors Suppliers Association, *Engineering Data Book*, FPS Version, Volumes I and II, Sections 1-26 (Tulsa, OK: 2004), 13-19.

External Heat Sources

"If you can't stand the heat, don't go to Cancun in the summer."

—Ben Stein [American writer, lawyer, actor, and commentator
on political and economic issues]

4.1 Introduction

The sun and *aerodynamic heating* are the two primary external heat sources affecting *aircraft thermal management* (ATM) during normal operations. *Lightning*, an infrequent (but regular) occurrence, is a third external heat source aircraft must accommodate.

Solar radiation affects both commercial and military aircraft by raising external structure temperatures, increasing vehicle heat loads, and reducing passenger and crew thermal comfort.

Aerodynamic heating also raises external structure temperatures as the kinetic energy of a moving vehicle changes to heat at the boundary air layer (boundary layer) in contact with the aircraft's external surface. Temperature increases above outside ambient are a thermal design challenge primarily for supersonic aircraft (Section 4.3.2).

Lightning (Section 4.4) can strike an aircraft on its way from the cloud to the ground during flight or while on ground. Each commercial aircraft faces about one lightning strike per year (on average), which provides a large threat to aircraft system performance, structural integrity, and passenger safety, with inadequate design measures.

4.2 Solar Heating

The sun affects more aircraft structure than any other heat source (at least on commercial airplanes), increasing some structure 100°F (38°C) or more above the *outside ambient temperature* (OAT). This makes accurate solar-heated structure temperature predictions a top priority in ensuring a minimum weight aircraft.

Solar heating increases the cabin cooling load, primarily through transmittance through windows. Larger passenger and flight deck windows on newer aircraft, such as the Boeing 787, increase the cabin cooling loads and allow more direct impingement on the passengers and flight deck crew compared to past airplanes.

The latest dimmable windows, which use a chemical that changes transparency to block the sun illumination (upon energizing with an electrical current) may also allow more heat into the cabin, compared to the white pull-down shades. The black surface of the dimmable window (which blocks light) is ideally suited for absorbing solar energy, which may end up as an increased cabin heat load. Black surfaces typically absorb 90% or more of the incident solar energy. The white pull-down shades, on the other hand, reflect most of the incident solar load striking them outward. White surfaces can absorb as little as 15% of incident solar energy, reflecting the remaining 85%.

Understanding the nature of solar heating and environmental and design factors affecting it influence an aircraft design, and passenger comfort is a critical part of ATM.

4.2.1 **The Sun**

The sun is a hot sphere of plasma that emits energy, through the process of nuclear fusion, as hydrogen isotopes fuse together to form helium under immense gravitational pressures and temperatures. Plasma is a fourth state of matter that consists of a cloud of charged particles (protons, neutrons, and electrons) created under high heat.

The fusion process generates huge amounts of energy from tiny amounts of matter, which scientists have been working to replicate (since the 1940s) to generate "clean energy." Fusion produces much shorter half-life radioactive by-products than fission, in addition to zero greenhouse gases. The massive amount of energy generated by fusion produces equally massive temperatures, estimated at 27 million °F (15 million °C) at the center of the sun [1].

The sun emits energy by electromagnetic *radiation* caused by the motion of charged particles in the material. The magnitude of the energy transferred is proportional to the absolute temperature of an object to the fourth power (Section 9.2.1.3). Fortunately for life on earth, the sun's outer core surrounds and absorbs almost all of the energy generated in the inner core, reradiating a tiny percentage of it outward at around 10,000°F (5800 K) [2].

FIGURE 4.1 The sun during total solar eclipse.

© SAE International

The sun also expels material generating a corona, which is only visible to the naked eye during a total solar eclipse, when the moon blocks the intense illumination from the surface of the sun (Figure 4.1).

The low density of the corona, shown in white (Figure 4.1), prevents the 2 million °F (1.1 million °C) temperature from significantly heating the planets.

The sun's radiation spans a range of *wavelengths* running from short waves consisting mostly of the *ultraviolet* (UV) (which degrades materials and human skin) and visible (which provides light) spectra to longer *infrared* (IR) waves (where heat transfer occurs). Solar radiation reaching the earth covers the spectrum seen in space minus atmospheric attenuation (Figure 4.2).

Under clear sky conditions, the energy of the sun reaching the earth comprises about 5% UV, 45% visible, and 50% IR [4].

FIGURE 4.2 Solar radiation spectrum reaching the earth's surface and atmosphere [3].

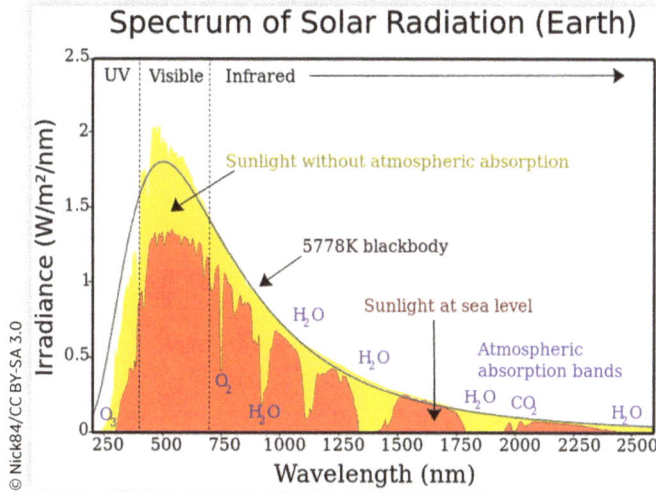

Spectrum of Solar Radiation (Earth)

While IR is associated with heat transfer, the absorption of UV and visible light also generates heat as their energy reradiates at a different temperature and corresponding wavelength. Solar energy that does not reradiate to the sky, or conduct to the colder ground below, heats the ambient air through convective heat transfer from the warmer ground. Solar heating is the primary cause of high ground OATs.

4.2.2 Incident Solar Load

The incident solar load reaching the earth varies with the season, location on earth, atmospheric conditions, and altitude. A maximum incident solar load occurs when the incident energy from the sun travels through the minimum atmosphere that lacks moisture or gaseous contaminants to absorb the energy.

Clear skies, with little moisture to attenuate the sun's rays, tend to occur at locations receiving little rain. These locations, including the Sahara Desert in Africa and Middle East, appear as parched earth on satellite photos (Figure 4.3).

While the African continent has few major airports along this dry land strip running north of the equator, some of the world's largest and busiest airlines operate from the Middle East. This makes it vitally important to design commercial aircraft to meet the environmental needs of a combined high solar load and OAT environment.

FIGURE 4.3 Satellite image of Africa.

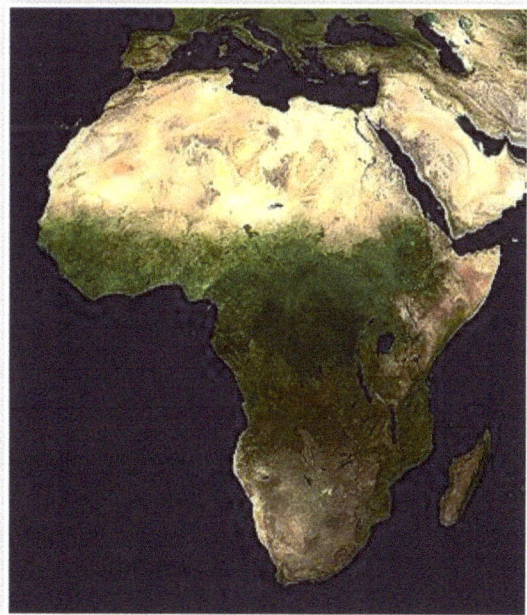

4.2.3 Time of Day and Surface Orientation

For horizontal surfaces, like the ground or an aircraft upper wing surface, the maximum solar loading occurs at *solar noon* when the sun is closest to being directly overhead.

Solar noon is the time of day when the sun's rays face the shortest path passing through the earth's atmosphere, generating the highest incident solar load for a *perpendicular surface*. At most locations on earth, solar noon varies from local noon because *local time* is based on 24 discrete time zones.

Peak daily solar loads occur on non-horizontal surfaces at times other than solar noon when the effect of a more perpendicular orientation to the sun's rays exceeds the greater atmospheric attenuation for a longer path through the atmosphere.

Solar reflections from the ground or adjacent aircraft surfaces can also increase the total solar load of structure beyond the maximum values reached by direct impingement. This is especially true as you move further down the skin of a circular fuselage to the lower lobe. Lower sun angles maximize the ground reflections to produce the peak structure design temperatures on the lower lobe.

The commonly referenced 355 Btu/h-ft^2 (1120 W/m^2) maximum ground solar flux (from *MIL-HDBK-310*) refers to a horizontal surface with no reflected solar load.

4.2.4 Solar Absorptance and Reflectance

Solar energy striking a surface is either absorbed, reflected, or (for transparent surfaces such as windows) transmitted. *Solar absorptivity* (α), *reflectivity* (ρ), and *transmissivity* (τ) are the fractions of solar energy absorbed, reflected, or transmitted by a surface, respectively. The sum of α, ρ, and τ equals one.

Solar absorptance is determined by the airplane surface optical properties, which can change in service from UV radiation exposure and contamination. Changes in α are well documented in spacecraft applications with extensive data sets showing *beginning of life* (BOL) and *end of life* (EOL) surface coating properties for satellites and spacecraft operating outside the atmosphere. Unfortunately, similar published datasets are unavailable in the public domain for aircraft. This is a major source of analytical uncertainty in aircraft external structure temperature predictions.

While orbiting satellites face a predictable environment of UV exposure, the dirt buildup on a wing and fuselage can vary widely between operational locations and airlines. With no manufacturer or certification requirements to clean aircraft surfaces, the pristine 0.20 α white wing at delivery can appear gray 10 years later. Especially, when operated by a fledgling airline with bigger concerns then washing their fleet. A surface takes on the α of the contaminate, proportional to the accumulated material. Dirt has a 0.9 α, so aircraft structure sized assuming temperatures corresponding to the initial 0.20 α for a white paint may provide inadequate margins in service. This is less of a concern with darker color paints having a higher α initially.

The α and ρ of the ground the aircraft sits on prior to the mission also affect structure temperatures by determining the amount of ground reflected solar heating. A dark surface, like asphalt, with a 0.90 to 0.95 α, reaches a higher temperature while reflecting less solar energy onto the structure compared to a lighter colored concrete. Concrete ρ values can vary widely as shown by Levinson and Akbari [5].

Energy efficiency standards have driven the development of higher ρ concrete to reduce ground surface (and air) temperatures and associated energy consumption for additional air conditioning. Aircraft thermal modeling must account for this trend, which can affect the latest runway construction in desert regions.

Running thermal models assuming both the maximum and minimum expected ground surface α ensures capture of the maximum structure design temperatures. This same issue of varying the surface α to obtain the maximum structure temperature also pertains to adjacent aircraft surfaces. A low α surface reflecting incident solar energy onto another aircraft surface with a high α would tend to generate the maximum temperatures for the high α surface.

4.2.5 Transmittance

Aircraft windows and external light enclosures transmit most, while also absorbing and reflecting some, of the solar energy striking them.

Aircraft passenger windows consist of multiple structural panes designed to maintain the cabin pressure and an inner acrylic window, called a dust cover, which protects them from passenger actions (Figure 4.4).

The gap between the structural windows provides a thermal resistance, like your home windows, which reduces heat transfer and the risk of fogging. On some newer commercial airplanes, an electrochemical panel replaces the inner surface plastic shade. A medium between two panes darkens upon applying a voltage, creating a dimmable window. The latest generation of dimmable windows can reduce the visible light transmittance to less than 0.0001% [6].

Flight deck windshields (Figure 4.5) consist of multiple layers laminated together to provide the strength needed to withstand a 4 lbm (1.8 kg) bird strike and survive thermal and aerodynamic loads.

Multiple structural panes are each designed to withstand all aircraft loads to provide redundancy. To prevent fog, the interlayer between the structural panes may also include embedded heaters, which can contribute to large thermal stresses in the windows. A heater design that fails to match the window heat to the in-service cooling variations over the glass surface can cause excessive thermal stresses leading to cracks. Manufacturing defects are another source of window failures.

Analysis methods developed to calculate the *solar heat gain* and *insulation values* for triple-paned residential and commercial buildings [8] are also appropriate for passenger windows.

Methods used for thermal modeling of spacecraft windows [9], meanwhile, are applicable to flight deck windows, upon adding external and internal convective heat transfer.

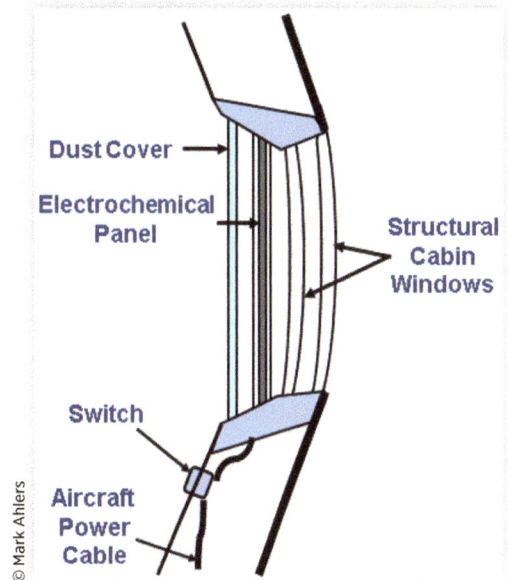

FIGURE 4.4   Dimmable passenger window.

Dust Cover

Electrochemical Panel

Structural Cabin Windows

Switch

Aircraft Power Cable

© Mark Ahlers

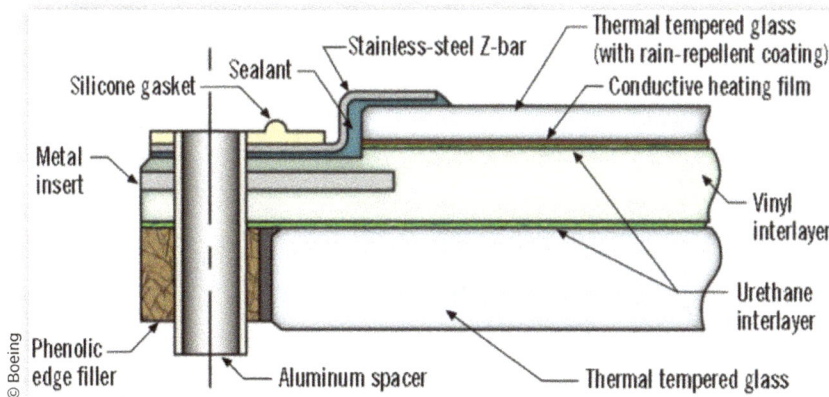

FIGURE 4.5   737 Flight deck windshields [7].

Silicone gasket

Sealant

Stainless-steel Z-bar

Thermal tempered glass (with rain-repellent coating)

Conductive heating film

Metal insert

Vinyl interlayer

Urethane interlayer

Phenolic edge filler

Aluminum spacer

Thermal tempered glass

© Boeing

4.2.6 Modeling Terrestrial Radiation

A confusing number of *clear sky models* (with varying levels of complexity) are available for predicting terrestrial solar irradiance. Most start with calculations to predict the *extraterrestrial solar* radiation and the light passage length through the atmosphere, based on a location on earth, date, and time. The simplest models end there, assuming atmospheric parameters corresponding to clear skies. Others include user inputs to vary atmospheric details, such as quantities of aerosols, moisture, dust, and clouds. The following equations support a simple clear sky model, which is extremely useful for checking results from computer program-generated fluxes, or building simple thermal models.

This analysis process starts with the *extraterrestrial radiation calculation* (I_0).

4.2.6.1 EXTRATERRESTRIAL RADIATION

The solar radiation emitted by the sun produces an average flux of about 433 Btu/h-ft^2 (1367 W/m^2) for a surface perpendicular to the sun at an average distance to the earth before atmospheric attenuation. This value is called the *solar constant* (I_{sc}).

The I_0 occurring just above the earth's atmosphere varies about 3.3 percent per year from the I_{sc} due to the elliptical orbit of the earth around the sun [10].

$$I_o \left\lfloor \frac{Btu}{hr-ft^2} \right\rfloor = I_{sc} \left(1 + 0.033 \cos\left(\frac{2\pi}{365} DOY \right) \right) \tag{4.1}$$

where

I_{sc} = solar constant,
DOY = day of the year, 1 = January 1st

4.2.6.2 DECLINATION ANGLE

The intensity of the sun is highly dependent on the length of the optical path through the atmosphere. This calculation starts with an understanding of the earth's seasonal orientation relative to the sun as defined by the *declination angle* (δ).

The sun's δ is the angle between the equator and a line drawn from the center of the earth to the center of the sun. It varies seasonally from plus 23.45° to minus 23.45° due to the earth's 23.45° tilt along its axis of rotation (Figure 4.6).

FIGURE 4.6 The earth's orientation versus season (Adapted from [11]).

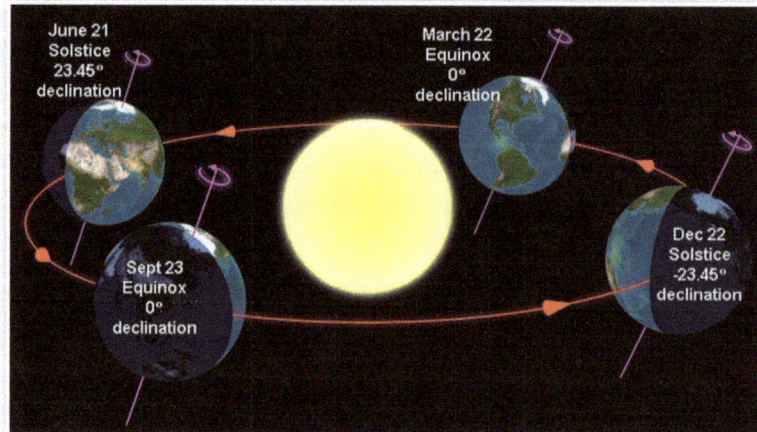

The earth rotates around its own axis every 24 h and the sun once every year. The summer solstice occurs on the day the sun appears furthest north or south in the sky for the northern and southern hemispheres, respectively. On the solstice, the sun appears directly overhead at *solar noon* along the latitude lines equal to the earth's tilt angle (23.45°). These latitude lines are the *Tropic of Capricorn* and *Tropic of Cancer* for the northern and southern hemispheres, respectively. The winter solstice occurs on the same day as the summer solstice on the opposite hemisphere. Equinoxes occur on the days when the sun first appears directly overhead on either hemisphere.

Peak solar loading, the most frequent design condition simulated, can occur anywhere between the Tropic of Cancer or Tropic of Capricorn by changing the date used to generate the solar flux.

The δ depends on the *DOY* [12].

$$\delta[\deg] = 23.45\sin\left(\frac{360°}{365}(DOY - 81)\right) \tag{4.2}$$

The 81 in Equation (4.2) corresponds to the Spring Equinox, the day when the sun appears directly overhead at the equator, leading to a 0° δ.

4.2.6.3 SOLAR TIME

After identifying the earth's location relative to the sun, a series of equations define the position of the sun in the sky with respect to the time of day.

Equation of Time. The *equation of time* (EoT) accounts for variations in the rate the sun crosses the sky during different times of the year. The eccentricity of the earth's orbit and the earth's axial tilt could add up to 16 min of error each year to the time equation without this correction [13].

$$EoT[\min] = 9.87\sin(2B) - 7.53\cos(B) - 1.5\sin(B) \tag{4.3}$$

where

$$B[\deg] = \frac{360}{365}(DOY - 81)$$

Local Standard Time Meridian. The *local standard time meridian* (*LSTM*) is a reference meridian running through the center of each time zone [14].

$$LSTM[\deg] = 15°\Delta t_{GMT} \tag{4.4}$$

Δt_{GMT} = the difference in local time from *Greenwich Mean Time* (GMT), h

Greenwich Mean Time is the yearly average (or mean) of the time when the sun crosses the Prime Meridian at the Royal Observatory in Greenwich, England. This was the first international standard time, which established a consistent 24-hour day across the globe. Today the GMT is also called the *Coordinated Universal Time*, or UTC (Table 4.1).

TABLE 4.1 Time difference from GMT in North America and South America time zones

Time zone	Δt_{GMT} (h)
Atlantic Standard Time	4
Eastern Standard Time	5
Central Standard Time	6
Mountain Standard Time	7
Pacific Standard Time	8
Alaska Standard Time	9

© SAE International

CHAPTER 4

Time Correction Factor. The *time correction factor* (*TC*) accounts for the time *variation* within a time zone [15].

$$TC\,[\min] = 4(Longitude - LSTM) + EoT \tag{4.5}$$

The number 4 in Equation (4.5) corresponds to the number of minutes it takes the earth to rotate 1°.

Local Solar Time. Twelve noon *local solar time* (*LST*) defines the time when the sun is highest in the sky.

$$LST\,[\text{h}] = LT + \frac{TC}{60} \tag{4.6}$$

where
    LT = local time, which usually differs from LST because of the earth's elliptical orbit, h

Hour Angle. The *hour angle* (*ω*) coverts the *LST* into an angular sun movement across the sky and equals 0° at *solar noon* [16].

$$\omega\,[\deg] = 15^{\circ}\,(LST - 12) \tag{4.7}$$

4.2.6.4 ZENITH ANGLE

The *zenith angle* (*z*), which is between the sun and the vertical, is the critical factor in determining the amount of atmosphere the sun must pass through to reach an aircraft (Figure 4.7) [17].

$$z\,[\deg] = \cos^{-1}[\cos\delta\,\cos\phi\,\cos\omega + \sin\phi\,\sin\delta] \tag{4.8}$$

where
    ϕ = latitude (+ for the northern hemisphere and − for the southern hemisphere), deg

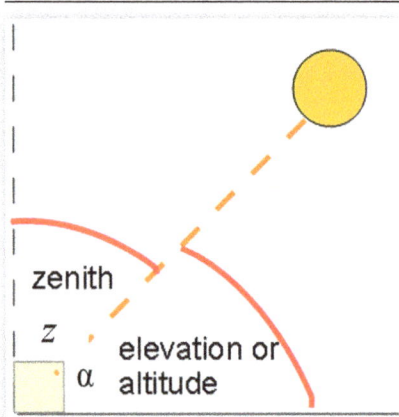

FIGURE 4.7 Zenith and elevation angles.

© SAE International

4.2.6.5 ALTITUDE OR ELEVATION ANGLE

The *altitude or elevation angle* (*α*), which is the angle of the sun measured from the horizon (Figure 4.7) is used to calculate the component of solar load striking non-horizontal surfaces.

$$\alpha\,[\deg] = 90^{\circ} - z \tag{4.9}$$

4.2.6.6 AIR MASS MODEL

The *air mass* (AM) model defines the optical path length light takes when passing through the atmosphere with respect to the minimum possible path occurring if the sun were directly overhead. Thus a 0° angle represents an AM of one while an AM of two occurs for a z around 60°. Equation (4.10) provides an AM correlation, which avoids divide-by-zero errors when z reaches 90° that occur for the commonly referenced 1/cos z [18].

$$AM = \frac{1}{\cos z + 0.50572(96.07995 - z)^{-1.6354}} \qquad (4.10)$$

4.2.6.7 CLEAR SKY MODEL

With the atmospheric path length identified, the *direct* and *diffuse* flux received by a surface are calculated.

Direct (or beam) irradiance comes directly from the sun, with *direct normal irradiance* (DNI) representing the flux received by a surface perpendicular to the sun's rays [19, 20].

$$DNI\left[\frac{\text{Btu}}{\text{h} - \text{ft}^2}\right] = 950.2\{1 - \exp[-0.075(90° - z)]\} \qquad (4.11)$$

Earth's atmosphere absorbs and scatters the non-direct radiation, producing diffuse irradiance in all directions. Diffuse irradiance also includes reflections from the ground, which depend on the *earth's ρ* (albedo) [21].

$$Diffuse\left[\frac{\text{Btu}}{\text{h} - \text{ft}^2}\right] = 14.29 + 21.04\left(\frac{\pi}{2} - z\frac{\pi}{180}\right) \qquad (4.12)$$

Global Horizontal Irradiance. *Global horizontal irradiance* (GHI) includes both the direct and diffuse radiation impinging on a horizontal surface [22].

$$GHI\left[\frac{\text{Btu}}{\text{h} - \text{ft}^2}\right] = DNI\cos(z) + Diffuse \qquad (4.13)$$

Global Angled Surface Irradiance. The *GHI* flux calculation is only useful for horizontal surfaces, which includes the ground and a small part of an aircraft. The corresponding flux for a tilted surface consists of a *direct* (S_{direct}) and *reflected* ($S_{reflected}$) component [23] (Figure 4.8).

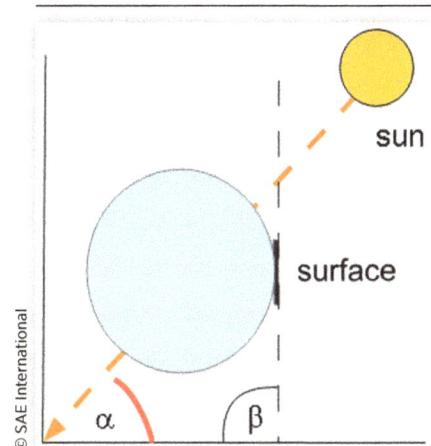

FIGURE 4.8 Solar radiation on tilted surface.

$$S_{direct}\left[\frac{\text{Btu}}{\text{h} - \text{ft}^2}\right] = DNI\sin(\alpha + \beta) \qquad (4.14)$$

$$S_{reflected}\left[\frac{\text{Btu}}{\text{hr} - \text{ft}^2}\right] = (DNI\sin\alpha + Diffuse)\rho\frac{(1 - \cos\beta)}{2} \qquad (4.15)$$

where
   B = angle of surface relative to ground, deg
   ρ = ground reflectance, fraction

Bird and Hulstrom [24] and Reno et al. [25] provide additional clear sky equations.

4.3 **Aerodynamic Heating**

Ambient air in contact with a moving aircraft skin takes on a higher adiabatic wall temperature as the air molecules decelerate compared to the free stream (Section 2.2.2.3). Convective heat transfer from the higher temperature air generates aerodynamic heating.

This heating effect is greatest at higher speeds that increase the outside ambient boundary temperature and convective heat transfer coefficients, and lower altitudes, where air density is greatest. It can have either a positive or negative impact on aircraft design and operation depending on the aircraft mission profile.

4.3.1 **Subsonic Flight**

For subsonic (below the speed of sound) flight, the ram temperature rise can have a large positive effect during high-altitude operation by keeping the airplane's skin, fuel, and hydraulic systems warmer. Airplane skin temperatures could dip below −112°F (−80°C) at higher altitudes, increasing structural fatigue loads without ram-air heating. Today's fuel and hydraulic fluids (which provide energy to move flight control surfaces) would also freeze at those temperatures.

At lower altitudes, where the ambient air temperatures can still reach relatively high values, ram-air heating can a have a negative impact on airplane operations. It reduces the level of cooling provided by ram-air systems removing waste heat from the airplane and the rate of solar-heated structural cooling from ambient air. Lower speeds (due to higher density air), however, limits aerodynamic heating at lower altitudes. This makes aerodynamic heating overwhelmingly positive for subsonic commercial airplanes.

Flat plate convective heat transfer correlations (Section 9.2.1.2) are used to calculate aerodynamic heating over much of an aircraft's external surface during subsonic flight.

4.3.2 **Supersonic Flight**

Unlike subsonic flight, the ram-air temperature rise is overwhelmingly negative for supersonic flight at all flight altitudes.

The retired British/French-built *supersonic transport* (SST) Concorde introduced the problem of aerodynamic heating to the commercial airplane market in the 1960s. Concerns over premature material degradation limited the Concorde's aluminum alloy skin temperature (Figure 4.9) to a peak value of 261°F (127°C). This corresponds to a maximum average cruise speed of around *Mach* (Ma) 2.0.

FIGURE 4.9   Concorde design cruise structural temperatures [26].

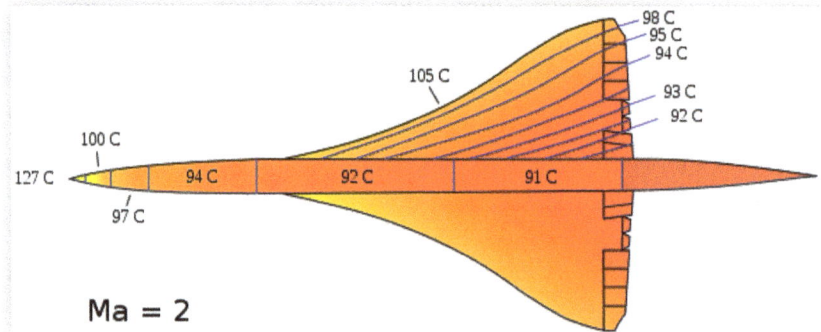

© Steel88/CC BY-SA 4.0

Aerodynamic heating caused the Concorde fuselage to expand up to one foot (300 cm) during a flight, creating fit issues within the cabin. The high adiabatic wall temperatures also forced the use of fuel as a heat sink for cabin cooling, similar to a jet fighter.

At the higher velocities of supersonic flight, aerodynamic heating requires a greater understanding of the aircraft boundary layer and a more involved analysis process, as defined in [27].

4.4 Lightning

Lightning is a very short duration electrical discharge within a cloud or between a cloud and ground, accompanied by a bright flash and usually a loud noise (thunder). It occurs after electrical charges build up within clouds due to particle collisions, and movement, similar to static electricity building up when you shuffle your feet while walking over a wool carpet. Stormy weather, with convective currents and ice and rain particles, increases the magnitude of the charge buildup, raising the risk of lightening. When the electrical potential between the cloud and ground exceeds some value, electrons surge from the clouds to the ground rapidly transferring many (large electrical currents) in pulses occurring over an extremely short time span. A single lightning bolt can contain up to 1 million volts at 30,000 A (for an extremely short duration) [28].

Lightning initially attaches to the airplane at one spot followed by pulses of electrons that attach further aft on the aircraft's skin before exiting from a series of locations (Figure 4.10).

In-flight lightning initially attaches at leading edge surfaces or sharp objects protruding in the airstream (i.e., antennae). This ionizes the air, generating a positive (electrical) charge that attracts negative charges in the cloud creating the lightning bolt.

Electrical particles stripped from the air in the process of ionizing allows them to flow through the airplane and to the ground, creating a lightning strike. The current running through the aircraft can wreak havoc on structure and systems lacking adequate protection or ignite fuel in the wing tanks (Section 8.12).

Permanent structural damage from high temperatures can occur if the electrical and thermal conductivity is too low to dissipate the energy. This problem intensified with the adoption of composite structures due to the greater electrical and thermal resistance relative to metallic structure.

FIGURE 4.10 Lightning path through an aircraft.

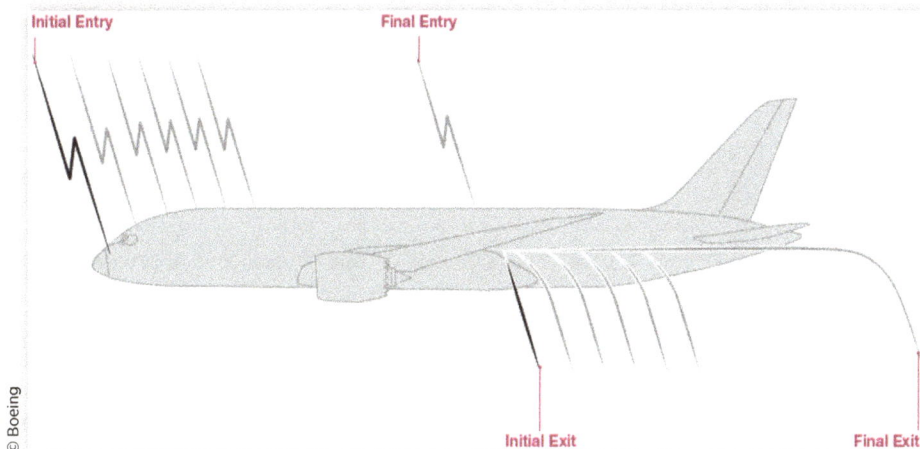

© Boeing

CHAPTER 4

References

1. European Joint Undertaking for ITER and the Development of Fusion Energy, "What Is Fusion," 2015, Fusion for Energy, accessed March 3, 2017, http://fusionforenergy. europa.eu/understandingfusion/.

2. Honsberg, C. and Bowden, S., "The Sun," PVEducation, accessed March 4, 2017, http://www.pveducation.org/pvcdrom/properties-of-Sunlight/the-Sun#footnote1_ h4izh3w.

3. Wikipedia, "Sunlight," 2017, accessed March 5, 2017, https://en.wikipedia.org/w/ index.php?title=Sunlight&oldid=782642238.

4. Wake, L.V., "Principles and Formulations of Solar Reflecting and Low Infrared Emitting Coatings for Defence Use," MRL-TR-89-2, Materials Research Laboratory, Commonwealth of Australia, 1989.

5. Levinson, R. and Akbari, H., "Effects of Composition and Exposure on the Solar Reflectance of Portland Cement Concrete," Lawrence Berkley National Laboratory, December 2001.

6. Driscoll, D., "Electronically Dimmable Aircraft Windows," *Aerospace & Defense Technology* (September 1, 2017), https://www.aerodefensetech.com/component/ content/article/adt/features/articles/27502.

7. Huff, J. "Improving the Service Life of Flight Deck Windshields," *Boeing Aero Magazine* no. 17 (January 2002), accessed March 5, 2017, http://www.boeing.com/ commercial/aeromagazine/aero_17/windshields.pdf.

8. Kee, M.S., Hughes, D.R., Lee, A.M., McMillan, M. et al., "Capstone Design: Energy Efficient Windows Triple Pane Window Analysis. S-2010," University of Tennessee Honors Thesis Project, May 7, 2010, 1–26, http://trace.tennessee.edu/cgi/viewcontent. cgi?article=2349&context=utk_chanhonoproj.

9. Luo, J.F., Yi, H.-L., Tan, H.-P., and Yang, L.-J., "Thermal Analysis of Optical Windows for Spacecraft Applications," *Journal of Thermophysics and Heat Transfer* 22, no. 2 (April–June 2008): 296–301.

10. Reno, M.J., Hansen, C.W., and Stein, J.S., "Global Horizontal Irradiance Clear Sky," Sandia Report SAND2012-2389, Springfield, VA: US Department of Commerce, March 2012,13, http://prod.sandia.gov/techlib/access-control.cgi/2012/122389.pdf.

11. Wikipedia, "Summer Solstice," 2017, accessed March 20, 2017, https://en.wikipedia. org/w/index.php?title=Summer_solstice&oldid=774653940/.

12. Mathew et al., 2005, 11.

13. Ibid., 11.

14. Ibid.

15. Ibid.

16. Ibid., 12.

17. Ibid.

18. Kasten, F. and Young, A.T. "Revised Optical Air-Mass Tables and Approximation Formula," *Applied Optics* 28 (November 15, 1989): 4735-4738.

19. Daneshyar, M., "Solar Radiation Statistics for Iran," *Solar Energy* 21 (1978): 345-349.

20. Paltridge, G.W. and Proctor, D. "Monthly Mean Solar Radiation Statistics for Australia," *Solar Energy* 18 (1976): 235-243.

21. Ibid.

22. Ibid.

23. ASHRAE, *2013 ASHRAE Handbook: Fundamentals (IP)* (ASHRAE, June 24, 2013), 14.11, ISBN-13: 978-1936504459.

24. Bird, R.E. and Hulstrom, R.L., "A Simplified Clear Sky Model for Direct and Diffuse Insolation on Horizontal Surfaces," Solar Energy Research Institute, SERI/TR-642-761, February 1981.

25. Reno, M.J., Hansen, C.W., and Stein, J.S., "Global Horizontal Irradiance Clear Sky Models: Implementation and Analysis," Sandia Report SAND2012-2389, March 2012.

26. Wikipedia, "Concorde," 2017, accessed January 12, 2017, https://en.wikipedia.org/w/index.php?title=Concorde&oldid=761703822.

27. Simsek, B., Kuran, B., Ali Ak, M., and Uslu, S. "Aerodynamic Heating Predictions Tool for a Supersonic Vehicle for Conceptual Design Phase," *AIAA Aviation, 46th AIAA Thermophysics Conference*, Washington, DC, June 13–17, 2016.

28. Sweers, G., Birch, B., and Gokcen, J., "Lightning Strikes: Protection, Inspection, and Repair," *Boeing AERO Online* QTR_04-12, 4th Quarter 2012, http://www.boeing.com/commercial/aeromagazine/articles/2012_q4/4/.

Aircraft Heat Sinks

Thank God men cannot fly, and lay waste the sky as well as the earth.

—Henry David Thoreau [US philosopher]

5.1 Introduction

Internally and externally generated heat (Chapters 3 and 4) must be removed from the aircraft to meet the structure, systems, and occupant temperature requirements (Chapter 2). Since energy can be neither created nor destroyed, the aircraft waste heat must be converted into a different form, or transferred to an external heat sink. While thermionic devices that convert heat to electricity exist, their extremely low efficiency, high weight, and high cost make them uneconomical for aircraft applications. Other efforts to use the waste heat to accomplish work, beyond heating cargo compartments and warming the fuel before burning it in the engines, have suffered a similar fate of excessive weight relative to the energy recovered. The waste heat low temperatures are the biggest hindrance to energy recovery. Consequently, in the near term transferring almost all waste heat to either *ambient air* or *fuel* is the only practical option for thermal management.

5.2 Ambient Air

Outside ambient air is the primary heat sink for commercial airplanes. It directly cools aircraft structure and equipment located outside the pressurized shell and removes heat from airplane systems to provide aircraft cooling. It also cools fuel in the main wing fuel tanks, increasing the capacity of the second primary heat sink.

5.2.1 Structure and Unpressurized Ambient Cooling

Free or *forced convection* to outside ambient air (and radiation to the sky and ground) cools external aircraft structure, which is warmer than the contacting outside ambient boundaries due to solar or aircraft system heating. The cooling effectiveness is proportional to the velocity of the airstream contacting the structure, ambient air density, and the adiabatic wall temperature (Section 2.2.2.3).

Still air on ground can generate convective cooling rates as low as 0.5 Btu/h-ft²-°F (2.5 W/m²-°C), over 30 times less than the values occurring during cruise for a commercial jet. This causes the 100°F (56°C) noon-hour temperature difference between the crown skin of a black fuselage and the outside ambient air seen during a sunny Phoenix summer day to drop to less than 5°F (3°C) 60 min later, while cruising at 36,000 ft (11,000 m).

Meanwhile, ambient air ventilates the unpressurized volumes during flight, cooling the enclosed equipment and systems. An uneven ram-air pressure distribution over multiple drainage and pressure equalization holes (on the external surface) drives airflow during flight. The flow is proportional to the airplane speed, like convective cooling of external structure.

Ambient ventilation is critical to cooling the *main landing gear* (MLG) brakes (Section 3.6) stowed in the wheel well during flight and maintaining an acceptable temperature for the surrounding structures and systems. Hot air emanating from the brakes could damage structure and systems without a constant flow of cooler outside ambient air to lower temperatures.

5.2.2 Systems Cooling

Compact *heat exchangers* (HXs) located in *ram-air systems*, or as a part of a *skin HX*, transfer aircraft systems-generated heat to the outside ambient air.

5.2.2.1 RAM-AIR SYSTEMS

Ram-air systems use the ram (total) air pressure (Section 2.2.3.2) in flight to drive outside ambient air through a compact HX, where it picks up aircraft waste heat (Figure 5.1).

An inlet designed to capture the maximum available ram pressure rise supplies a duct that directs the cooler outside ambient air through a HX and out an exit. A fan drives the outside ambient air during ground operations, making this a much more effective heat transfer method compared to a skin HX.

Compact Heat Exchangers by Kays and London [1] provides a thorough coverage of HXs designed to meet the low weight and volume aircraft requirements.

A movable entrance door controls the flow and louvers control the direction of the flow exiting the system on some systems to reduce the drag associated with taking boundary layer air. The door matches the cooling flow rates to the system needs (by changing the flow area), while the exit louvers control the exit air direction to maximize

FIGURE 5.1 Ram-air system.

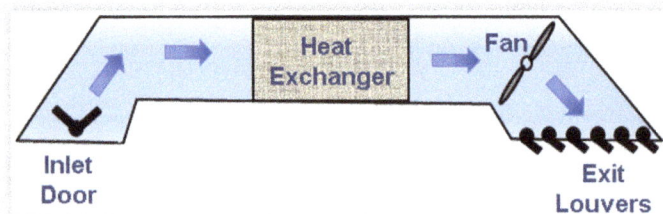

© SAE International

the thrust recovery. *Air-conditioning* (AC) (Section 7.2.2) packs are the biggest ram-air users on commercial transport aircraft.

Ram-air systems are much more widely used than skin HXs on aircraft because they transfer more heat to the ambient environment using a smaller volume, with fewer design constraints, while weighing less.

5.2.2.2 SKIN HEAT EXCHANGERS

Skin HXs reject heat to the outside environment using the airplane's external surface instead of drawing outside ambient air through an air-to-air or liquid-to-air HX. The high convective heat transfer coefficients combined with low outside-air temperatures make skin HXs a potentially appealing option for removing aircraft heat loads occurring during flight on subsonic aircraft. Terrible ground performance, with low external air speeds and associated convective heat transfer coefficients, and high boundary air temperatures limit skin HX applications. A 5 mph (8 km/h) wind speed during ground static operation, or even worse still air, can cause the external convective heat transfer coefficients to plummet. Meanwhile, a –22°F (–30°C) standard day adiabatic wall (air) temperature at 36,000 ft (11,000 m) can turn into a 113°F (40°C) OAT during ground operations, when airplane cooling needs are usually greatest.

Locating sufficient skin area to accommodate the contacting fluid channels of the HX core is the biggest design challenge to skin HXs. Composite fuselages have increased the problem of finding adequate surface area since the lower thermal conductivity severely reduces the heat transfer from a hot fluid flowing along an airplane's skin, eliminating its potential use.

5.2.2.3 CABIN EXHAUST

Air exiting the pressurized volume at a higher temperature than the outside ambient also removes cabin heat. On large commercial airplanes, this includes airflow through an *outflow valve* (OFV) designed to maintain the cabin pressure (Section 7.2.5), or *electrical/electronic* (EE) cooling system (Section 7.4.1) *overboard exhaust valve* (OEV). Even uncontrolled leakage through door seals and fuselage penetrations removes cabin heat.

Maximum heat removal occurs by directing the warmest cabin air (usually air heated by equipment) toward dedicated exhaust paths. This cooling approach can, however, have negative consequences during certain operating scenarios. OFVs can almost completely close during descent for bleed-air pressurization systems, as low engine power settings reduce the pressure available to run the AC packs and provide air to pressurize the cabin. Hot air blown at a mostly closed OFV can overwhelm the exhaust capacity causing reverse airflow from the lower lobe to the main deck. This may allow heated air from the lower lobe to reach the passenger cabin.

5.3 Sky

Deep space also acts as a heat sink for aircraft on ground and in flight, with a much lower relative effect on ATM compared to air. An aircraft sitting on an airport tarmac during on overcast day could see a temperature a mere 18°F (10°C) cooler then OAT. Although, during static operations on ground, the cooling effect can be large enough, relative to convective heat transfer, to justify modeling. The effect on aircraft temperatures is negligible, however, relative to convective cooling for in-flight vehicles.

5.4 Fuel

A fully loaded wing fuel tank contains a large thermal capacitance. Commercial and military aircraft use this thermal capacitance as a heat sink for highly variable aircraft heat loads generated by hydraulic systems. Military aircraft frequently depend on fuel to also cool electronics and weapon systems in addition to hydraulics to offset the lack of ram-air systems. Stealth requirements limit the use of ram-air systems, with their hot exhaust air to attract heat-seeking missiles.

5.4.1 Thermal Capacitance

On commercial airplanes, hydraulic systems generate high heat loads while raising or lowering landing gears and doors or rapidly moving control surfaces to counteract air turbulence. Long periods of greatly reduced hydraulic actuation and heat generation usually follow these short duration events. Hydraulic fluid often bypasses the fuel tank HXs to avoid overcooling during periods of low heat generation.

Burning fuel decreases the fuel tank thermal capacitance, which cooling system designs must account for. Anticipating at the start of each flight the effect of a decreasing thermal mass combined with reduced cooling, due to a shrinking wing wetted surface area, ensures an adequate fuel sink for the planned mission. High fuel temperatures, due to inadequate fuel relative to the heat load, combined with high outside ambient temperatures, is a concern for its effect on *fuel flammability* and the fuel cooling system performance.

Higher fuel temperatures increase the risk of fuel tank vapors (ullage) igniting (Section 6.2.6), causing an explosion in the presence of a sufficiently high energy source such as an arcing wire. Larger commercial transport (30 passengers or more) and military aircraft include inerting systems to reduce this flammability risk by lowering the ullage oxygen content (Section 8.9.2.2).

5.4.2 Fuel Supply Line and Energy Recovery

For larger more consistent heat loads, which would overwhelm the cooling capacity of a tank of fuel over a long mission, the engine fuel supply line provides a second option for energy recovery. A HX transfers airplane-generated heat to the fuel stream supplying the engines. Inputting waste heat into the engine fuel feed line raises the fuel temperature before it enters the engine, reusing 100% of the energy. This provides one of the few economic options for energy harvesting (on an aircraft).

The engines recover some of the energy of waste heat entering the bulk fuel (from the hydraulics), which continually transfers heat to the colder outside ambient environment. The engine oil cooler uses the engine fuel supply as a heat sink because engine waste heat generation and fuel flow both follow increases in engine thrust levels. Using this fuel stream as a heat sink also adds to the criticality of the bulk fuel temperature since higher temperatures reduce the cooling capacity of the fuel entering the engine. High fuel temperatures, which provide inadequate cooling for the engine oil, can reduce the engine performance or life due to elevated component temperatures.

Reference

1. Kays, W.M. and London, A.L., *Compact Heat Exchangers* (Malabar, FL: Krieger, 1998), ISBN-10: 1575240602.

Fires and Failures

"Success consists of going from failure to failure without loss of enthusiasm."

—Winston Churchill [British statesman, novelist]

6.1 Introduction

The *Aviation Safety Network* (ASN) reported zero commercial airplane passenger deaths from crashes of scheduled jets in 2017 [1]. While the ASN also reported 44 occupant and 35 ground fatalities from crashes of turboprops, freighters, and unscheduled charter operators [2], the aviation safety record still far exceeds other transportation modes. Those 79 deaths occurred in a year when more than 3.7 billion passengers flew commercial airplanes.

This impressive passenger safety record is due in large part to aircraft designs that ensure *continued safe flight and landing* (CSFL) following failure conditions. Commercial airplane manufacturers must demonstrate system and structure designs with an extremely improbable chance of one or multiple failures causing a catastrophic hull loss. Extremely improbable is defined as one catastrophic hull loss occurring no more than once every one billion (1,000,000,000) flight hours. This United States (US) *Federal Aviation Administration* (FAA) certification requirement, which refers to a *10-to-the-minus-9* event, drives commercial airplane safety. The FAA is the most influential of the regulatory agencies for different countries, which must approve all aspects of the commercial transport aircraft industry effecting passenger safety. This ranges from the initial airplane design to the production and maintenance facilities and procedures.

Since multiple systems and structural components can fail, calculating the probability of a hull loss failure is more difficult. Although, the Dutch aviation consulting firm To70 estimated the fatal accident rate at one for every 16 million large commercial airplane passenger flights based on actual crashes [3].

Thermally related failures include *fires* and system *overheat conditions* producing excessive temperatures. A *cargo fire*, which heats adjacent aircraft structure, or a *dragging brake* during takeoff, which overheats the *main landing gear* (MLG) wheel well following gear retraction, are typical thermal-related failures.

6.2 Fires

A fire (especially in flight) is one of the most hazardous events that can occur in an aircraft. Smoke can reduce visibility for the pilots, making it difficult to see their instrument panels to control the airplane or engage the systems designed to fight the fire. Burns and noxious fumes can incapacitate the crew and passengers. Meanwhile, critical structure and flight systems can fail due to exposure to extremely high temperatures, ultimately leading to a crash. The 1600 to 2000°F (871 to 1093°C) flame temperature, representing a typical aircraft fire in FAA testing (Section 11.4), far exceeds the temperature capability of aircraft structure and systems, excluding the engines and *auxiliary power unit* (APU).

Government regulatory agencies, such as the FAA in the USA and *European Union Aviation Safety Agency* (EASA) in Europe, mandate fire detection and control requirements designed to prevent catastrophic aircraft failures leading to a loss of life.

Automated fire detection systems are a certification requirement for the MLG wheel well, engine, APU, and cargo compartments on commercial transports. The engine and APU must also include systems to extinguish fires, while cargo compartments use systems to suppress (rather than extinguish) fires.

6.2.1 MLG Wheel Well

MLG wheel well fires normally occur due to sloppy or faulty maintenance that allows flammable materials to contact hot wheel surfaces [4]. Excessive or incorrect grease usage in lubricating the wheels or cleaning solvents and hydraulic fluids left on brake heat shields can all ignite during landing when the brakes reach their maximum temperature. Heat shields limit radiative heat transfer from brake surfaces to aircraft structure or systems.

While these fires normally do not cause major damage or jeopardize passenger safety, a few catastrophic fires occurring (on older aircraft designs) during takeoff have also caused the death of all on board. A more recent incident occurred in 1991 on a Nigeria Airways' McDonnell Douglas DC-8 (built in 1968), which crashed shortly after takeoff from Saudi Arabia. Underinflated tires appeared to initiate a series of events leading to an in-flight fire and crash, killing all 261 occupants onboard.

When *nitrogen* (N_2) was not readily available to fill the tires, ground personnel released the airplane (allowed the airplane to leave the gate and fly) with underpressurized tires rather than risk a delay. Nitrogen, an inert gas, replaces the combustible *oxygen* (O_2) in air-filled tires to reduce the fire risk from an exploding tire. The low tire pressure caused excess deflections and stresses in the adjacent tire during taxi, leading to a failure during the takeoff roll. Soon after, the adjacent tire failed as it took on the load of the failed tire. A wheel then locked up, generating even higher temperatures, which ignited the tire. While the crew knew something was wrong during the takeoff roll (after the tire blew), the aircraft lacked a system to indicate high temperatures and crew procedures to extend the MLG. Retracting the landing gear into the wheel well allowed the fire to consume the passenger cabin floor, which is directly overhead, causing a decompression event that collapsed more of the floor structure. Multiple passenger seats and strapped-in passengers soon fell through the hole in the floor structure and out the airplane.

The loss of life from this type of crash drove the FAA to mandate fire detection systems that alert the crew to take corrective measures if a fire occurs in a MLG wheel well.

High air temperatures measured close to the MLG pressure deck, usually with thermocouples or a continuous wire or eutectic loop, provide fire detection on modern commercial airplanes. A temperature increase in the wire or melting of the eutectic material changes the electrical resistance of the loop to indicate a high temperature associated with a fire or hot brake. Running an electric wire or eutectic loop back and forth across the pressure deck provides more thorough coverage compared to point measurements provided by a few thermocouples.

Upon fire detection, the flight deck personnel are instructed to take corrective action, which usually means lowering the gear as soon as the airplane is at a safe speed and altitude. This moves the fire threat away from airplane structure, while letting the freestream air extinguish the flames, like a child blowing out candles on their birthday cake.

During the time required for fire detection, crew reaction, and possibly dropping to the correct altitude and speed, the fire can heat structure and systems far beyond normal design temperatures. Therefore, flight-critical structure and systems may need thermal-related mitigation in addition to the fire detection system. The mitigation can include covering structure with fire resistance materials or oversizing adjacent structure to carry additional loads from fire-damaged structure.

While MLG wheel well fires are extremely rare for commercial airplanes, overheated brakes are likely to occur during the life of the airplane. The fire detection system, therefore, must differentiate between a fire scenario and hot brake event. Brake temperature measurements are also available to warn the crew of excess temperatures that can overheat structure or equipment or start a fire.

6.2.2 Engine and APU

The engines and APU on large commercial airplanes incorporate both automatic fire detection and extinguishing systems. The detection system typically senses both fires and high temperatures using detector loops or point locations, like the system in a MLG wheel well.

Aircraft manufacturers also design means to reduce the risk and severity of fires including

- Minimizing ignition sources

- Routing flammable fluid-carrying lines away from electrical wires and hot pneumatic ducts (ignition sources)

- Compartmentation to isolate fuel leaks and ignition sources (Figure 6.1)

- Firewalls to prevent a fire from spreading

- Shut-off means for flammable fluids into and out of the fire zone

- Drainage and ventilation to minimize the accumulation of flammable fluids and vapors

The fire-extinguishing system rapidly releases a fire-extinguishing agent, currently Halon, from pressurized bottles following fire detection. Halon smothers the flames, replacing O_2-carrying air feeding the flames. The APU fire detection-extinguishing system follows a similar approach.

Future airplane programs are banned from using Halon because it contributes to depletion of the ozone layer, leading to a search for an environmentally and economically acceptable replacement.

FIGURE 6.1 Engine compartmentation [5].

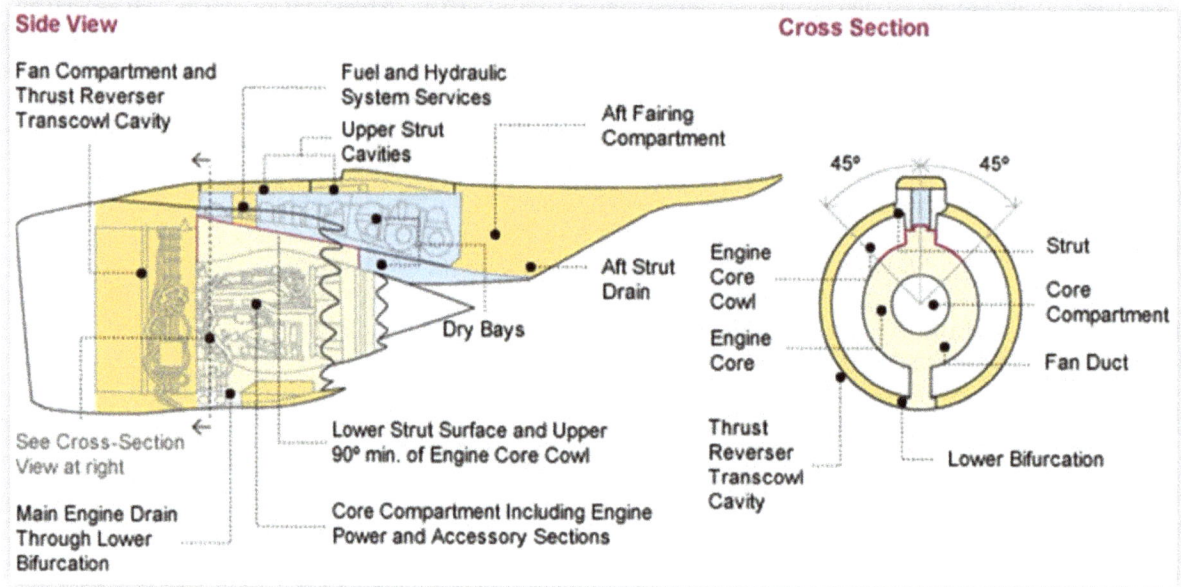

Side View

Fan Compartment and Thrust Reverser Transcowl Cavity

Fuel and Hydraulic System Services

Upper Strut Cavities

Aft Fairing Compartment

Cross Section

45°   45°

Engine Core Cowl

Engine Core

Aft Strut Drain

Dry Bays

Strut

Core Compartment

Fan Duct

See Cross-Section View at right

Main Engine Drain Through Lower Bifurcation

Lower Strut Surface and Upper 90° min. of Engine Core Cowl

Core Compartment Including Engine Power and Accessory Sections

Thrust Reverser Transcowl Cavity

Lower Bifurcation

© Boeing

6.2.3 Cargo Compartment

Commercial aircraft cargo contains a wide range of combustible materials (Table 6.1). Passenger aircraft carry cargo in lower lobe compartments while freighters carry cargo in both lower lobe compartments and the main deck. The FAA established a classification system for cargo compartments, in federal regulation *14 CFR 25.857*, based on the method used to detect and fight a fire occurring within them.

TABLE 6.1 Permitted cargo combustible materials.

Permitted cargo combustible materials		
Rubbing alcohol	Perfume	Colognes
Hair spray	Nail polish remover	Paper
Alcoholic beverages	Small arms ammunitions	Lithium-ion batteries

© SAE International

6.2.3.1 CLASS A

In class A cargo compartments, a flight crew member at their station can easily discover a fire and the entire compartment is easily accessible in flight. The flight crew, therefore, acts as a fire detection and extinguishing system.

6.2.3.2 CLASS B

In a class B cargo compartment, a crew member can reach a fire occurring anywhere within the compartment with an extinguisher without stepping in. No hazardous quantities of smoke, flames, or extinguishing agent must enter any compartments occupied by crew or passengers, and there is an approved smoke or fire detector to warn the crew. The crew acts like the extinguishing system while the fire detection is automated.

6.2.3.3 CLASS C

Class C cargo compartments have approved built-in smoke or fire detector systems and fire-extinguishing or suppression systems. They include means to exclude hazardous quantities of smoke, flames, and extinguishing agents from areas occupied by passengers or crew, and controls to limit ventilation within the compartment, to allow the suppressant agent to perform as designed. Suppressant agents reduce the O_2 content enough to prevent visible flames.

Class C compartments are required for the lower lobe of commercial airplanes that are inaccessible to the crew during flight, which covers all present-day passenger aircraft.

6.2.3.4 CLASS E

Class E cargo compartment include a system to detect a fire and means to stop ventilating airflow from entering and prevent hazardous quantities of smoke, flames, or noxious gases from reaching the flight crew compartment.

These compartments allow the fire to extinguish themselves by consuming all compartment O_2.

6.2.3.5 CLASS F

Class F cargo compartments are located on the main deck and have fire detection and suppression systems and means to prevent hazardous quantities of smoke, flames, or extinguishing agents from reaching passenger or crew areas. Class C compartments have the same requirement.

Fire Detection and Suppression Systems. The fire detection and suppression systems are similar to systems used on the engine and APU with a few differences. First, upon fire detection for some airplanes, the release of the fire suppressant is dependent on a flight crew action. This adds a delay that allows the fire to grow, or the possibility of the fight crew ignoring the alarm. The second major difference is that the initial large release of an agent, which eliminates visible flames, may not extinguish the fire. Therefore, a continuous (but much reduced) fire suppressant flow rate soon follows.

Flames are extremely hazardous due to temperatures and convective heat transfer rates high enough to melt metals. The continuous supply of suppressant gas offsets the effects of air leakage into the compartment to maintain a sufficiently low O_2 concentration in the air to prevent flame growth. Therefore, cargo fires can smolder until the airplane lands at the closest airport, which may be hours away. Upon opening the cargo door (following the evacuation of the airplane), the smoldering fire can turn into a raging inferno as air entering the compartment feeds O_2-starved embers.

Heat from the smoldering fire during flight can raise the surrounding structural temperatures (beyond the flame impingement area) above the normal design operating temperatures, defining another structural design condition.

Freighter aircraft, which do not carry passengers, have their own cargo compartment fire protections requirements referred to as classes E and F.

6.2.3.6 CARGO LINERS

The different cargo compartment types share a common liner requirement. Liners must prevent flame penetration while subjected to flame impingement for 5 min during FAA required burnthrough testing (Section 11.4.1.3). In-service experience and extensive testing has shown that meeting this requirement ensures adequate containment to support CSFL to an airport following an "expected" fire event. Therefore, if passengers or shippers avoid sneaking hazardous and prohibited items onboard, like filled O_2 bottles, the safety risk from the rare cargo fire is well controlled.

6.2.4 Passenger and Crew Area

Measures designed to *prevent*, *detect*, and *extinguish* fires minimize their risk in the passenger and crew areas.

6.2.4.1 PREVENTION

Prevention is the first line of defense with ignition sources, such as electrical wires and hot ducts, identified and isolated from combustibles. The FAA also requires the use of fire protective materials inside the cabin that are self-extinguishing (stop burning) upon flame removal.

6.2.4.2 FIRE DETECTION

Smoke detectors installed in the lavatories, crew rest compartments, and in some crew work areas, like galley complexes and purser work stations, provide fire detection. Occupied areas in the main cabin depend on the passengers and cabin crew to see smoke.

6.2.4.3 FIRE SUPPRESSION

Commercial airplanes include built-in fire extinguishers for disposal receptacles in lavatories. The lavatories are designed to contain a fire for 30 min, which allows time for the fire to safely consume the fuel source if the cabin crew are unable to gain access.

Handheld fire extinguishers are located throughout the airplane cabin for manual firefighting.

6.2.5 Electrical/Electronic Bay and Lower Lobe

An abundance of ignition sources, between power feeders and electrical equipment, fill *electrical/electronic* (EE) bays. Aircraft manufacturers use different designs for handling EE bay fires.

Boeing's 787 has a smoke detector to indicate a fire, which warns the flight crew to switch to the smoke override mode which sends smoke-filled air directly overboard through the closest *outflow valve* (OFV).

Embraer, the Brazilian manufacturer of regional (smaller passenger) jets (which was purchased by Boeing), uses a fire detection and suppression system for the EE bay on their KC-390 Tanker transport. They treat their EE bay like a cargo compartment, minimizing air exchange to allow a suppressant agent to maintain a low enough O_2 level to starve flames from a fire [6].

Concerns over EE bay fires revolved around electrical short circuits igniting a limited amount of dust and debris prior to the failure of lithium-ion batteries on 787 aircraft during January of 2013.

6.2.5.1 787 LITHIUM-ION BATTERY FIRES

On January 7, 2013, about a minute after all passengers and crew members disembarked Japan Airlines Flight 008 at Boston's Logan Airport, a cleaning crew spotted smoke in the aft cabin of the 787-8. A lithium-ion battery undergoing thermal runaway with one failed cell causing intense heat and smoke which spread to the remaining seven cells caused the event. The battery consists of eight cells, tightly packed in an aluminum box.

Lithium-ion batteries bring new challenges to EE bay fire protection by producing a much more rapid and severe thermal effect during their failure compared to past electrical fires.

The US *National Transportation Safety Board* (NTSB) investigation concluded that that a short-circuit in the 787 battery was the likely cause for the first cell overheating, while inadequate thermal isolation between cells allowed the remaining cells to overheat. It took the rapid release of energy from multiple cells failing one after another and the splattering of electrolyte beyond the confines of the battery box to create the charred surfaces of Figure 6.2.

FIGURE 6.2 Aft electronics bay after a burning battery in a Boeing 787 [7].

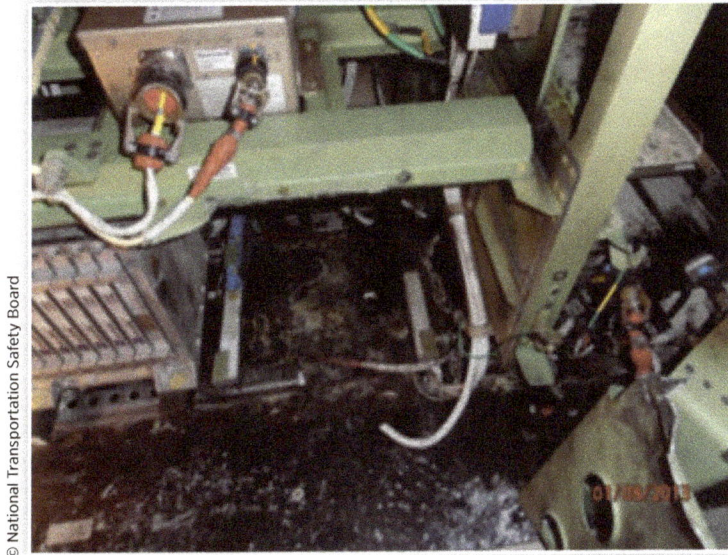

© National Transportation Safety Board

The level of media coverage for this incident was remarkable considering the relative risk to passenger safety. Critical *system separation requirements* ensured that any flight critical equipment affected by the heat or electrolyte release had redundant systems to assume that functionality if needed. This requirement is designed to ensure CSFL following detonation of an explosion device disabling all aircraft systems within a 6 ft (1.83 m) diameter sphere.

During a second battery event (in-flight), the smoke alarm warned the crew to switch the air distribution system to the smoke override mode, turning off recirculation fans to prevent combustion by-products from reaching the main cabin. The recirculation fans are part of a system adding filtered cabin air to the supply to provide additional cabin ventilation (Section 7.2.3.1). Meanwhile, air drawn through electronic boxes located in the EE bay (to provide cooling) was blown toward an *outflow valve* (OFV) sending battery-generated smoke overboard (Section 7.4.1).

An improved battery thermal design that limited the failure to a single cell would have dramatically reduced the peak temperature of the released gases. It might have even been possible to design the battery box to contain the electrolyte from one failed cell. Better quality control in the battery manufacturing process and an improved thermal design should have prevented these failures from occurring.

The battery manufacturer responded by providing additional thermal isolation between cells in the battery (under Boeing's guidance) and plans for improved quality control. Meanwhile, Boeing developed a containment system to isolate and exhaust the by-products of a battery failure safely overboard. It consists of an extremely rugged metal box connected to a vent tube penetrating the airplane's skin.

Today's more stable (safer) battery chemistries, compared to the vintage 2005 787 design, would likely reduce the risk of a similar failure on the next commercial aircraft.

For those who look for the benefit from negative events, the in-flight 787 battery incident demonstrated the effectiveness of aircraft designs in ensuring CSFL following a fire event in the EE bay. Fire procedures worked as planned for more severe thermal conditions than anyone anticipated occurring in an EE bay. The copious amounts of combustion gases vented overboard and systems beyond the battery continued operating normally.

6.2.5.2 WHY LITHIUM ION?

The 787 electrical system uses lithium-ion batteries for the same reason as electric cars, laptop, tablets, and cell phones. They carry a much higher energy density compared to the traditional nickel-cadmium battery and can recharge much quicker. They also require less maintenance, with no need to completely discharge them between recharge cycles (to avoid a memory effect).

The main battery operates for brief periods on ground, powering aircraft systems prior to engine start or after engine shutdown, while the aircraft is connected to a pushback tractor. The APU battery supplies power to start the APU, which can start the airplane engines. Both the main and APU batteries can provide emergency power during flight in the event of an extremely rare power failure.

Generators on the engines generate electrical power to recharge the batteries and run airplane systems.

The major disadvantages to lithium-ion batteries are a greater susceptibility to failing if they get too warm or a short circuit occurs, and more severe heating and smoke generation. Higher power density and charging rates accompany an impressive rapid discharge. Scary high energy release rates can make traditional firefighting chemicals and methods impotent.

The *International Air Transport Association* (IATA), the major airline trade association, released regulations in January of 2017 limiting the charge state of lithium-ion cells shipped in aircraft to 30% of their maximum capacity, to reduce the fire risk. Lithium battery fires in a cargo compartment are a much bigger risk to passenger safety than the 787 battery fires, with abundant material available to ignite in the immediate vicinity.

6.2.6 Fuel Tank Fires

On July 17, 1996, a 747-100, Trans World Airlines (TWA) Flight 800, crashed into the ocean shortly after takeoff from New York City's John F. Kennedy International Airport. An explosion, fire, and impact with the ocean surface destroyed the airplane (Figure 6.3), killing all 230 passengers and crew.

FIGURE 6.3   TWA Flight 800 recovered aircraft parts [8].

© National Transportation Safety Board

The NTSB concluded that a spark from short-circuiting wires leading to the *fuel quantity indicating system* (FQIS), located inside a fuel tank, most likely caused the explosion. A contributing factor to the explosion was high ullage (air/fuel vapor) temperatures in the *center wing* fuel *tank* (CWT). A one-hour delayed takeoff provided more time for the three *air-conditioning* (AC) packs, mounted directly below an "empty" CWT, to heat the fuel and ullage, raising the temperature and flammability risk. An empty 747-100 CWT contains a shallow puddle of about 50 gallons of fuel that pumps supplying the engine are unable to scavenge. This provided plenty of fuel to fill the tank with vapors but too little to provide a significant thermal lag under extended pack heating, which created a flammable fuel/vapor mixture.

New requirements to reduce the risk of an explosion, including the installation of *nitrogen-generating systems* (NGS) (Section 8.9.2.2) to reduce O_2 concentrations in fuel tanks, resulted from the accident investigation.

6.2.7 External Fuel Fire

External fires following a crash can occur when fuel from a ruptured fuel tank or line ignites upon contacting hot engine surfaces. The extremely lightweight insulation blankets mounted on the fuselage skin, which protect passengers from external ambient temperatures and outside noise, must also provide thermal protection from high-temperature flames. A radiant panel tests ensures that the blanket covering materials provide more time for passengers to escape when flames from an external fuel fire threaten the fuselage.

6.3 System Failures

Some airplane systems contain pressurized fluids at elevated temperatures running through ducts and tubes that can develop leaks. The leak size can range from a catastrophic *burst duct* releasing massive amounts of energy to a pin hole expelling a small amount of air, which goes unnoticed.

6.3.1 Burst Ducts

Bleed-air ducting transfers pneumatic energy from the engine, in the form of hot pressurized air, with the potential to damage surrounding airplane structure and systems during a duct failure. Seals connecting duct sections to valves or other equipment are the most common failure points. Although, ducts can also burst far away from seals initially releasing 110 lbm/min (50 kg/min) or more of 480°F (250°C) air at 40 psi (275 kPa) [9].

$$\text{For a 400-passenger airplane } 0.55\frac{\text{lbm}}{\text{occupant}-\text{min}}*200\frac{\text{occupants}}{\text{airplane}}=110\frac{\text{lbm}}{\text{min}}$$

A rapid pressure rise in the constrained volume of a leading edge could cause a panel blowout, jeopardizing airplane control. Meanwhile, high-temperature air can damage front spar structure, aircraft wiring, and flight critical equipment.

Equipment in the pack bay and adjacent structure is also vulnerable to a duct burst, with high-pressure and -temperature ducting feeding the AC packs located directly below a fuel tank on many commercial airplanes.

6.3.2 Leaking Ducts

While internal system monitoring will usually quickly detect catastrophic duct failures (from pressure changes in the ducting system), smaller leaks continually heating adjacent structure and equipment can be difficult to detect. This lower leakage rate is primarily a concern if it negatively affects aircraft structure or systems.

The same wire or eutectic loop designs used in the MLG wheel, engine, and APU compartments can also detect duct leaks and burst events, following changes to accommodate different temperature set points.

References

1. Shepardson, D., "2017 Safest Year on Record for Commercial Passenger Air Travel: Groups," *Reuters*, January 1, 2018, https://www.reuters.com/article/us-aviation-safety/2017-safest-year-on-record-for-commercial-passenger-air-travel-groups-idUSKBN1EQ17L.

2. Reed, D., "Nobody Died in Commercial Jet Crashes In 2017: Good News, But Not As Good As You Might Think," *Forbes*, January 3, 2018, https://www.forbes.com/sites/danielreed/2018/01/03/nobody-died-in-commercial-jet-crashes-in-2017-good-news-but-not-as-good-as-you-might-think/#115b48021644.

3. Shepardson 2018.

4. Weber, B., "Preventing Wheel/Brake-Area Fires," *Boeing Aero_Qtr_2007*, February 2007, http://www.boeing.com/commercial/aeromagazine/articles/qtr_2_07/article_04_1.htmlBoeing.com.

5. Hariram, S., Paul, P., and Dummeyer, D., "Fire Protection Engines and Auxiliary Power Units," *Boeing Aero*, QTR_04.10, 2010, http://www.boeing.com/commercial/aeromagazine/articles/2010_q4/3/.

6. Air Force Technology, "Embraer Selects Meggitt's Fire Protection System for Brazilian KC-390," August 21, 2012, http://www.airforce-technology.com/news/newsembraer-selects-meggitts-fire-protection-system-for-brazilian-kc-390.

7. Hersman, D.A.P., "Investigative Update of Battery Fire on Japan Airlines B-787," National Transportation Safety Board, January 7, 2013, https://www.skybrary.aero/bookshelf/books/2079.pdf.

8. National Transportation Safety Board, "Aircraft Accident Report, In-flight Breakup Over the Atlantic Ocean Trans World Airlines Flight 800," Boeing 747-131, N93119, Near East Moriches, New York, July 17, 1996, Accident Report NTSB/AAR-00/03,2000, Washington, DC: Government Printing Office, https://www.ntsb.gov/investigations/AccidentReports/Reports/AAR0003.pdf.

9. Flight Safety Foundation, "Bleed Air Leaks," 2017, Skybrary Aviation Safety, Last modified September 21, 2017, accessed Oct 20, 2017, https://www.skybrary.aero/index.php/Bleed_Air_Leaks.

CHAPTER 7

Environmental Control Systems

"It was luxuries like air conditioning that brought down the Roman Empire. With air conditioning their windows were shut, they couldn't hear the barbarians coming."

—**Garrison Keillor [American author, storyteller, humorist]**

7.1 Introduction

Environmental control systems (ECS) perform multiple functions to *sustain life*, ensure *passenger comfort* and *safety*, and *airplane performance* during operations in hostile conditions. They provide conditioned air to pressurize the cabin and control temperatures; chill food and beverages; remove heat from the electrical equipment; detect, extinguish, or suppress fires; prevent or remove ice from external surfaces and sensors; and exhaust odors from lavatories and galleys.

This chapter introduces the multiple ECS affecting *aircraft thermal management* (ATM).

7.2 Cabin Temperature and Pressure Control

Cabin temperature and pressure control makes up the largest portion of the ECS functionality and consumes the most energy on a commercial vehicle, after propulsion. It involves pressurizing and cooling the outside ambient air to provide a comfortable environment for airplane occupants.

Figure 7.1 shows the ECS responsible for providing cabin pressure and temperature control.

The engine (1) compresses outside ambient air, which flows through the bleed-air system to the ozone converter (2) just prior to entering the *air-conditioning* (AC) packs (3). The AC packs cool the air stream, which mixes with recirculated and filtered cabin

FIGURE 7.1 ECS responsible for cabin temperature and pressure control.

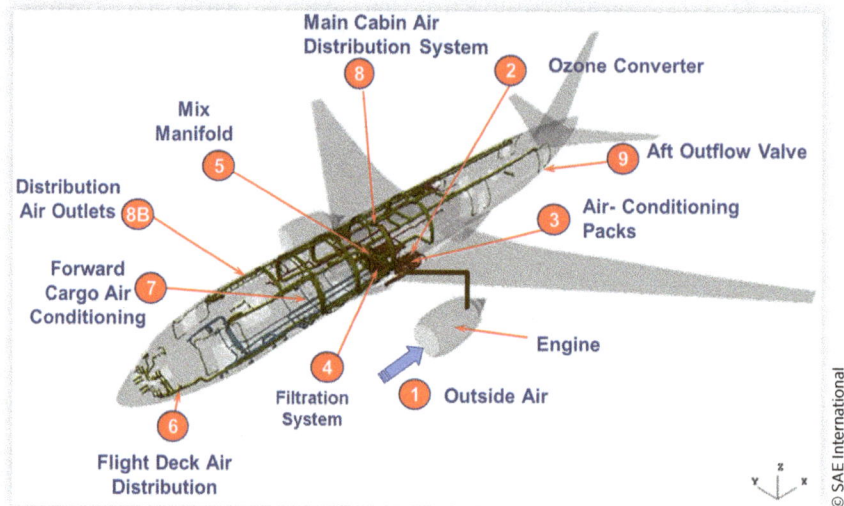

exhaust air in the mix manifold (5). Either mixed (outside and recirculated) or unmixed (outside only) temperature-controlled air flows to the flight deck air distribution system (6), depending on the aircraft model. Some of the mixed air may also supply a forward cargo AC system (7), while the majority supplies the air distribution system (8), which includes overhead air outlets (8B) designed to control flow patterns in the passenger cabin. One or two *outflow valves* (OFVs) (9) control the cabin pressure by metering the exhaust flow. Following is a more detailed description of each section.

7.2.1 Air Supply (1 to 2)

A bleed-air (supply) system distributes high-pressure, high-temperature air bled from an engine to support airplane systems such as cabin pressurization and AC, wing and engine inlet ice protection, hydraulic reservoir pressurization, hydraulic pumps and flap drives power generation, *nitrogen-generating system* (NGS) operation, potable water pressurization, and *total air temperature* (TAT) probe aspiration.

Modern turbofan jet engines provide propulsive motion by pushing large amounts of air backward at a high speed. The engine compresses outside ambient air in multiple stages before reaching the combustor where adding fuel ignites the air fuel mixture. This creates a high-pressure gas that expands in a turbine powering a fan blowing *bypass* air around the compressor. The fan generated bypass air provides aircraft propulsive power. Aircraft systems can extract compressed air from bleed ports upstream of the combustor (Figure 7.2) to meet airplane system needs.

Engines compress outside ambient air to provide sufficient *oxygen* (O_2) (during flight) to support combustion in the engines, and life in the cabin, given the reduced outside ambient air pressure. Air pressure matters because it defines the amount of available O_2 for breathing and combustion. To improve the engine operating efficiency, the pressure of ambient air supplied to the combustion chamber is much higher than required to ignite the fuel. Higher air pressures, which cause higher air temperatures, help reduce the fuel burn for each new engine.

Air supplied at higher pressures than needed to pressurize the cabin can provide power for AC packs and other airplane systems. AC packs convert pressure to mechanical

© Boeing

FIGURE 7.2 Typical turbofan engine.

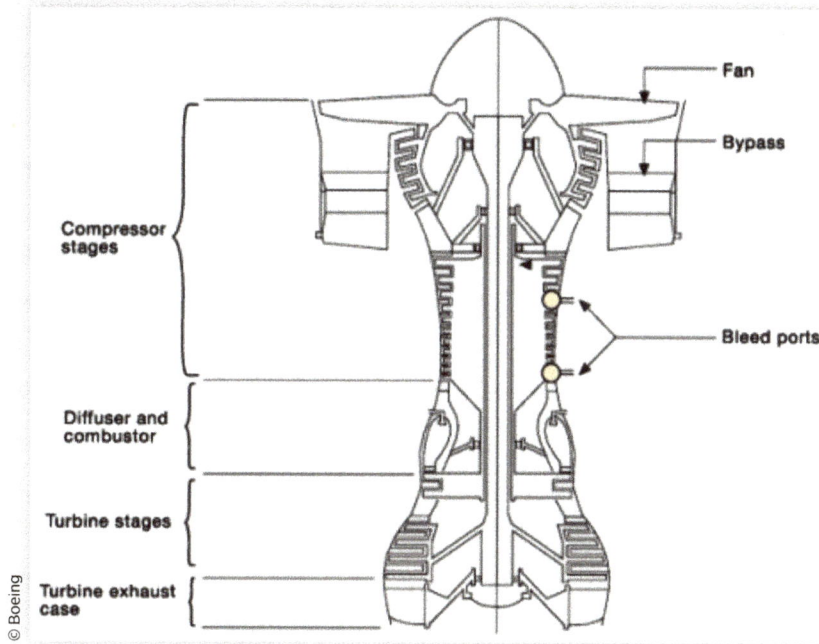

power by running bleed air through turbines connected to compressors supporting a refrigeration process (Section 7.2.2).

Engine settings and ambient pressure variations during a mission make multiple engine bleed ports necessary for the extracted air pressure to more closely match airplane system requirements. Designers trade increases in engine cost and weight for additional bleed ports against the added fuel burn from providing more excess pressure from fewer ports.

A *heat exchanger* (HX) (called a precooler) transfers some of the heat generated in the compression process to outside ambient cooling air supplied by the engine-driven fan air (Figure 7.3). Bleed air exiting the precooler is still normally a toasty 350°F–400°F (177°C–205°C), which usually exceeds subsonic aircraft structural design temperatures (except for the engine and *auxiliary power unit* [APU]).

FIGURE 7.3 Engine bleed-air system.

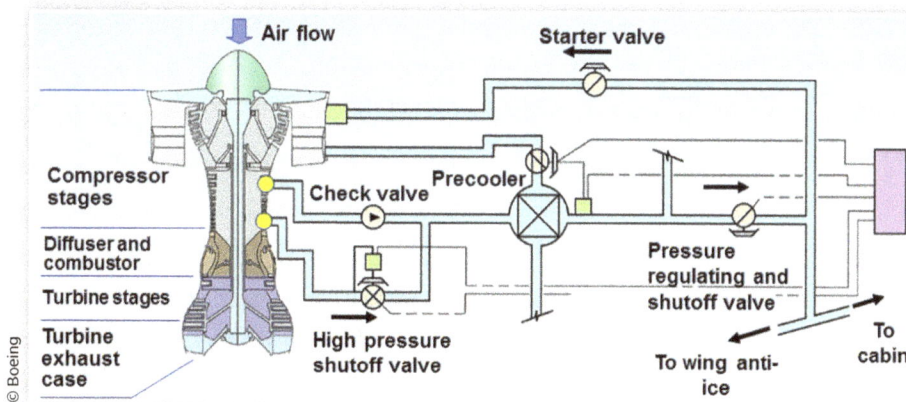

© Boeing

The United States (US) code of federal regulations *14 CFR 25.831 (a.)* requires the air supply system to provide a minimum of 0.55 lbm/min (0.25 kg/min) of outside air per occupant during normal operations. The 0.55 lbm/min is equivalent to 10 cfm (0.283 m³/min) at an 8000 ft (2438 m) cabin altitude, the maximum allowed by the US *Federal Aviation Administration* (FAA).

The flight deck receives more air (on a per occupant basis), which provides a positive (higher) pressure relative to adjacent compartments while meeting greater cooling requirements. The positive pressure prevents the inflow from adjacent compartments, preventing smoke or gases from entering the flight deck through exhaust grills or door gaps. The additional air also provides cooling to offset large equipment and solar heating loads.

Most propulsive power comes from engine fan bypass air. Higher bypass engines burn less fuel for the same propulsive affect by bypassing more air around the core for each gallon of fuel burned.

As bypass ratios increase (Figure 7.4) extracting bleed air for non-propulsive uses, such as pressurizing a cabin, has a greater impact on engine performance. If this trend of higher bypass airflow continues, eventually bleed-air systems may no longer be a viable option for the latest generation of energy efficient commercial jet engines. This would force the widespread adoption of *no-bleed systems*.

FIGURE 7.4 Commercial transport engine bypass ratios (Data from [1]).

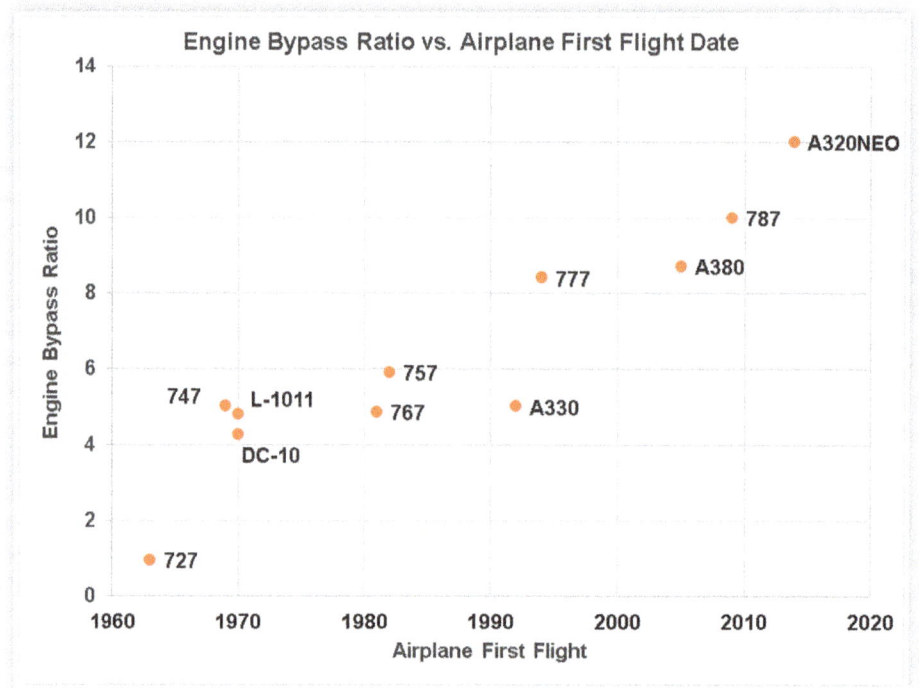

© SAE International

7.2.1.1 NO-BLEED SYSTEM

Cabin air compressors (CACs) powered by the rotating engine shaft or electricity produced by engine-mounted generators pressurize outside ambient air in no-bleed systems. With this design, the pressure generated by the compressor can drive the AC packs, similar to a bleed-air system. Airplane system functions, such as wing thermal protection and cabin cooling, can also be electrically driven, reducing the need for engine-generated high-pressure air.

7.2.1.2 GROUND-BASED OPERATION

During ground static operations, the APU located in the airplane tail or a ground cart, can provide high-pressure air to supply pneumatic power or run AC packs. An APU is a mini-engine designed to provide electrical power and pressurized air, while a ground cart is a more efficient APU on wheels. APU design focuses on minimum weight and exceptional reliability because they operate a small percentage of the time and provide flight critical functionality, such as starting engines during in-flight shutdowns.

Airlines can also provide conditioned air downstream of the pack with ground carts supplying temperature-controlled air. This provides more energy efficient cooling, extends the life of the pack, and reduces the heating of aircraft structure and fuel for airplanes with packs located beneath the *center wing* fuel *tank* (CWT). Higher fuel tank ullage vapor temperatures increase the flammability risk.

7.2.2 Air Conditioning (3)

Pressurized air supplied by the engine bleed system or CACs is hot after the compression process and requires cooling before entering the cabin, where it provides temperature and pressure control. On the ground and shortly after takeoff, the air temperature required for cabin cooling is often lower than the outside ambient air. Thus, heat transfer from a lower temperature source to a higher temperature sink occurs using a compression refrigeration cycle. Compression refrigeration uses the physical phenomena of compressed fluids getting colder as they expand to achieve low air temperatures using either an *air cycle machine* (ACM) or *vapor cycle machine* (VCM).

7.2.2.1 AIR CYCLE MACHINE

Most turbine-powered aircraft use ACMs to cool the high-pressure and high-temperature air [2].

ACMs use high-pressure air as both the power source and refrigerant in a reverse Brayton cycle, which transfers the heat of compression from the engine bleed air to the outside ambient. Commercial airplanes typically have two ACMs (Figure 7.5) to prevent a total loss of cooling if one fails.

FIGURE 7.5   Two ACMs.

© Boeing

FIGURE 7.6   4-Wheel bootstrap ACM.

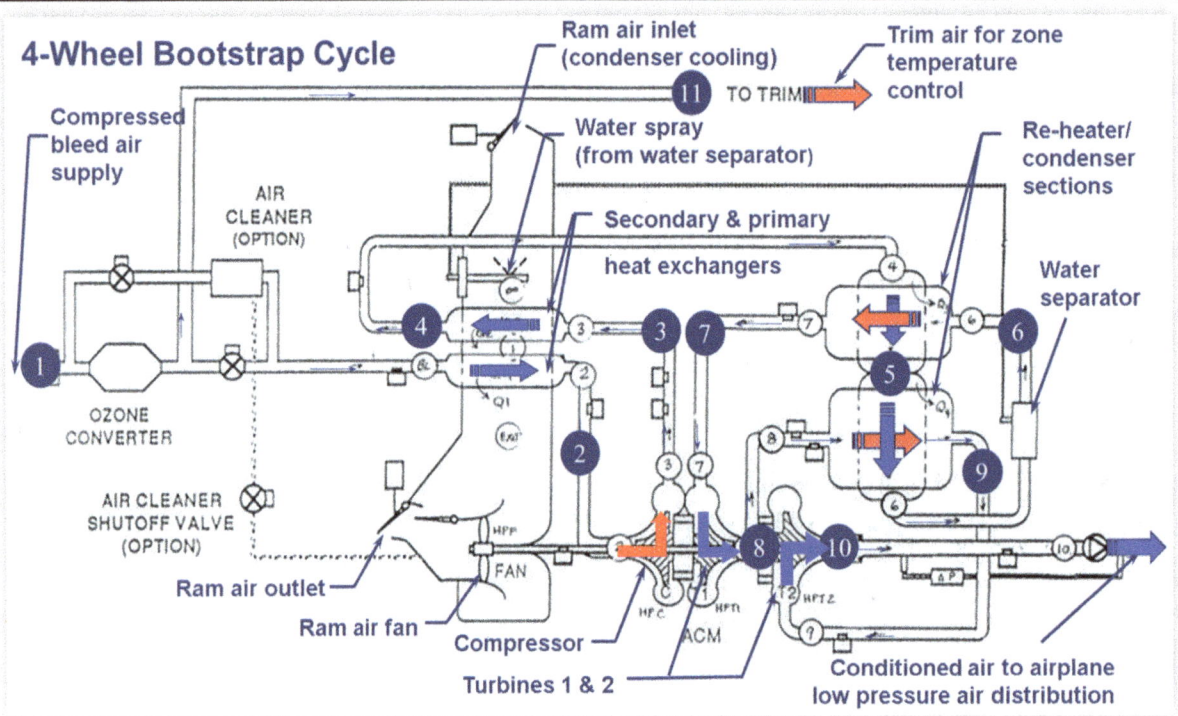

4-Wheel Bootstrap Cycle

© SAE International

"Packs," an abbreviation for *package*, is a commonly used term for ACMs.

An AC pack includes two HXs cooled by ram (outside) air, a compressor, and one or more turbines. The ram-air duct inlet captures air flowing over the airplane's skin (ram air) in flight, while a fan provides cooling air to the HXs during ground operations.

On most commercial jets, the wing-to-body fairing underneath the center wing of the airplane's structure contain the AC packs in a *pack bay*. The overhead structure serves as a fuel tank on many larger aircraft making the pack bay a *flammable leakage zone*, since fuel tanks can leak. Special precautions in flammable leakage zones ensure that no surfaces in contact with fuel vapors or liquid reach a high enough temperature to support *autoignition*. Safety requirements include prohibiting materials that can absorb fuel in flammable leakage zones, areas where fuel can collect.

Figure 7.6 shows a typical commercial aircraft 4-wheel bootstrap cycle ACM.

The four wheels refer to the ram-air fan, compressor, and two turbines, while bootstrap indicates their placement on the same shaft. The shared shaft makes the energy removed in expanding higher-pressure air in both turbines (to lower the air temperature) available for compressing the air (earlier in the cycle) and powering the ram-air-cooling fan during ground operations. A second turbine provides cooler supply air temperatures to meet greater cabin cooling requirements. Aircraft also fly with ACMs using different numbers of wheels (components), like a 3-wheel machine with one turbine.

ACM Thermal Process. Compressed, pre-cooled, bleed air supplying the ozone converter (1) during flight passes through the primary HX, which rejects heat to the outside air. *Ozone* (O_3) convertors convert O_3 molecules found in atmospheric air at higher cruise altitudes, to O_2. Ozone is a lung irritant and health hazard at high enough levels.

On ground a fan provides outside ambient cooling air to ECS HXs, and in-flight the ram-air pressure rise drives the flow (Section 2.2.2.3).

A small amount of hot air (called trim) is removed (11), prior to passing through the primary HX, and used to vary (or trim up) the temperature of conditioned air supplied to the different cabin zones. Independently controlled temperatures zones accommodate nonuniform heating and cooling loads occurring along the cabin. Cabins with a wide variation in seating densities have similarly uneven cooling loads.

Cooled air exiting the primary HX (2) passes through the first wheel of the ACM, a compressor (2 to 3), which increases its pressure and temperature. The higher pressure makes it possible to expand the air further inside the turbine (later in the cycle) to reach a lower temperature. The higher temperature also increases the amount of heat transferred to the outside air by raising the temperature difference.

The compressed supply air passes through the secondary HX (3 to 4) where it rejects heat to the ram cooling circuit (outside air). The air then passes to the re-heater section (4 to 5), where the heat it rejects warms the air coming from the condenser section prior to entering the first-stage turbine (7). Warming the condenser outlet air helps ensure that water in the air stream, which the water separator fails to capture (6), evaporates before entering the first-stage turbine.

The air continues to the condenser section (5), where it again rejects heat, thus condensing moisture. A water separator (6) extracts some of the moisture, which supplies a water spray to lower the outside air heat sink temperature using evaporative cooling during ground and low-altitude operations. The air then travels thru the re-heater (6 to 7), where the secondary HX outlet air adds heat to evaporate entrained liquid in the airstream prior to passing thru the "second wheel" of the ACM, the first-stage turbine (7 to 8) where air expands.

Cooled air from the first-stage turbine outlet (8) passes back through the condenser section (8 to 9), where it acts as a heat sink, while re-heated air from the condenser section travels to the second turbine (third wheel) of the ACM (9).

The air then expands in the second-stage turbine (9 to 10), decreasing the pressure and temperature, while the common shaft absorbs the energy from the expanding fluid. The conditioned (refrigerated) air then mixes with recirculated cabin air and hot trim air, to satisfy the range of heating and cooling needs.

ACM Temperature versus Specific Entropy Diagram. A *temperature versus specific entropy* (T-s) diagram provides a useful means for analyzing the efficiency of an AC thermodynamic cycle. *Specific entropy* (*s*) represents energy over temperature. Figure 7.7 shows a T-s diagram for the ideal Figure 7.6 ACM architecture.

The T-s diagram provides a visual representation of energy transfers in the individual thermodynamic processes of an AC pack, as well as the overall system efficiency. Vertical lines occur when the refrigerant, which is air for Figure 7.7, expands or compresses, while the angled lines represent heat transfer. The area under the curve is equal to the heat transfer. While this example assumes a 100% efficient process in which all energy generated from the air expansion in the turbines (5 to 8 and 9 to 10) is available to compress the air (2 to 3), in real-life conversion, losses release part of the available energy as waste heat. Inefficiencies (energy losses) appear as tilted vertical lines, as the entropy increases during the compression and expansion processes.

Different areas of the T-s diagram are proportional to the net cooling provided and corresponding power requirement for the cycle. In Figure 7.7, the area enclosed by the lines represents the work input while the area underneath the graph lower lines, labeled *Cooling Load*, captures the ACM cooling. A primary goal of pack design is to minimize the area enclosed by the lines while maximizing the area underneath the lower curve.

FIGURE 7.7   4-Wheel bootstrap ACM temperature versus entropy diagram.

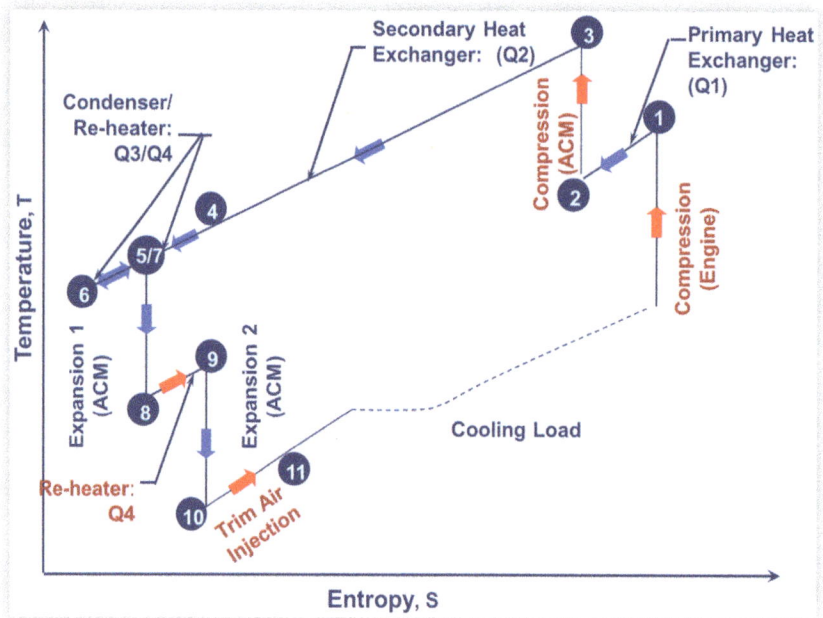

You will note that the tilt of the compression and expansion lines increase the power consumption, but not the cooling production.

7.2.2.2 VAPOR CYCLE MACHINE

Less commonly seen on large commercial jets are VCMs, which operate similarly to a refrigerator using a two-phase refrigerant to transfer heat in a closed loop cycle. Figure 7.8 shows a simplified VCM.

FIGURE 7.8   Vapor cycle machine (Adapted from [3]).

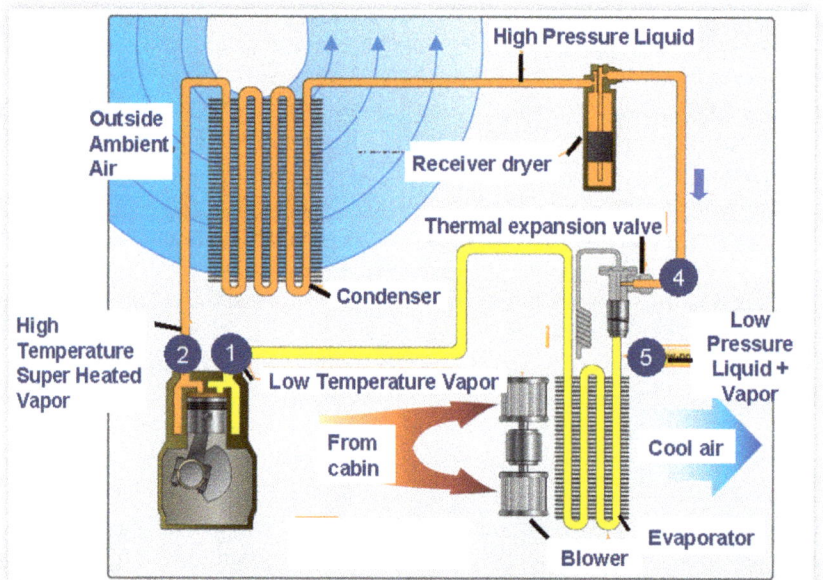

VCM Thermal Process. In a vapor cycle, heat transfer occurs due to a phase change in a two-phase refrigerant releasing more energy, compared to compressing or expanding air, for the same input work. A VCM includes a compressor and condenser, similar to an ACM, while a thermal expansion valve reduces the refrigerant pressure instead of a turbine. The VCM also includes low- and high-pressure refrigerant in the liquid and vapor phases, unlike the ACM cycle which is entirely gaseous.

A liquid reservoir, called a receiver, stores the refrigerant under pressure. The liquid flows from the receiver through a thermal expansion value (4 to 5) reducing the pressure to allow the fluid to evaporate as it picks up heat from the cabin supply air. The refrigerant passes through a compressor (1 to 2) increasing the vapor pressure and temperature creating a superheated vapor, which rejects heat to the outside ambient air in a condenser (2). The vapor condenses to a saturated liquid after heat removal before flowing back into the receiver.

VCM T-s Diagram. The VCM T-s diagram (Figure 7.9) looks different compared to the ACM due to the nature of the two-phase heat transfer.

A constant temperature line representing heat transfer occurring in the condenser (3 to 4) and evaporator (3 to 4) is the most obvious difference compared to an ACM. The net cooling and work input are, however, represented in the same way, using areas underneath the curve.

FIGURE 7.9   VCM T-s diagram [4].

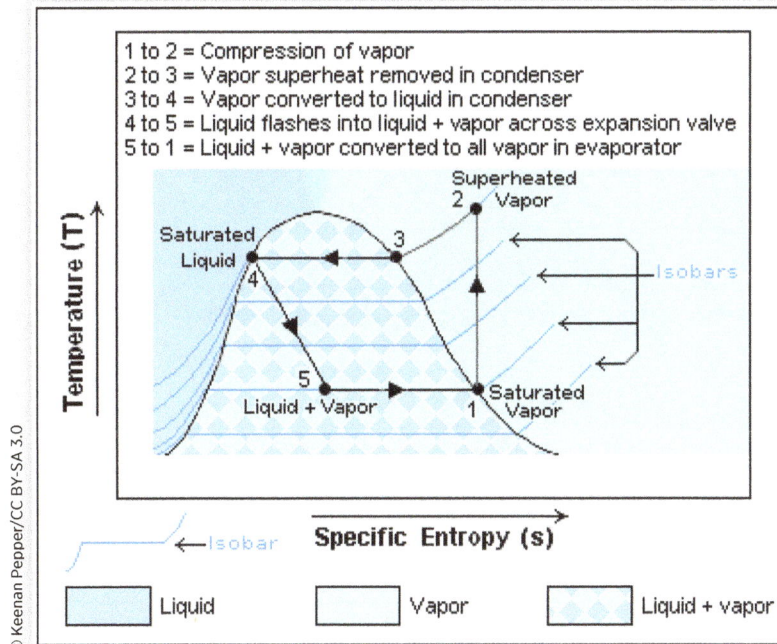

© Keenan Pepper/CC BY-SA 3.0

7.2.2.3 AIR CYCLE VERSUS VAPOR CYCLE MACHINE

ACMs have been the preferred cooling system for airplanes with ready sources of high-pressure air for years because they are lighter, more reliable, require less maintenance, and avoid the negative consequences of refrigerant leaks associated with VCMs. Their low efficiency and dependence on high-pressure air sources, however, could lead to a resurgence in VCMs, which are more energy efficient and better suited for more electric aircraft operations. CACs and vapor cycle cooling provided cabin pressurization and temperature control for the Boeing 707 in the 1950s. Concerns over contaminants in the bleed air was the major driver back then, which is normally not an issue with modern jetliners. Poor reliability and high weight led later commercial jets to adopt ACMs.

7.2.3 Air Distribution (4-8)

The air distribution system supplies conditioned air in a way which ensures acceptable thermal comfort and adequate contaminant removal throughout the cabin. Key parts of the air distribution system include the *recirculation*, *main cabin*, and *flight deck air distribution systems*.

7.2.3.1 RECIRCULATION SYSTEM

Commercial airplanes include recirculation systems which reuse some of the cabin's exhaust air to ventilate the cabin. The typical system combines pack outlet air with recirculated cabin air captured by fans supplying a *mix manifold* where both air streams mix. Mixed air exiting the mix manifold supplies the air distribution system (Figure 7.10).

FIGURE 7.10 Recirculation system.

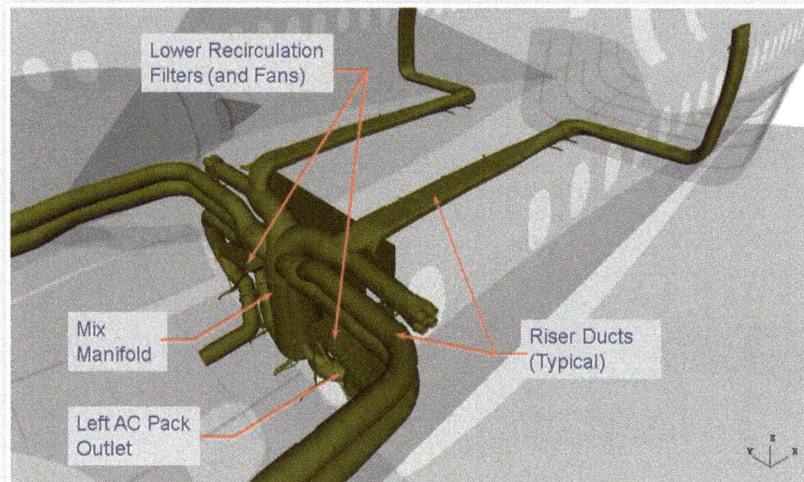

Filters clean recirculated cabin air entering the mix manifold. Large commercial airplanes use hospital-quality *high-efficiency particulate arrestor* (HEPA) filters that remove a minimum of 99.97% of bacteria and airborne particles that viruses use for transport [5].

All modern commercial transport aircraft use recirculated air, usually about 50%, to save the energy required to pressurize and cool additional outside air needed beyond meeting ventilation requirements. This additional flow supports cabin cooling using a more comfortable overhead nozzle supply temperature, and sufficient air velocities to cool the occupants and prevent reverse flow from the lower lobe.

Recirculating cabin air also raises the cabin humidity, which provides a more comfortable breathing environment. Outside air at cruise altitudes contains almost no moisture. Recirculating cabin air, which includes moisture transpired from passengers, is, therefore, a key contributor to preventing the cabin air from becoming even dryer.

7.2.3.2 MAIN CABIN

The main cabin air distribution system supplies the passenger cabin, flight deck, personal air outlets (traditionally called gaspers), lavatories and galleys, crew rest, and door areas (Figure 7.11).

FIGURE 7.11 Main cabin air distribution system.

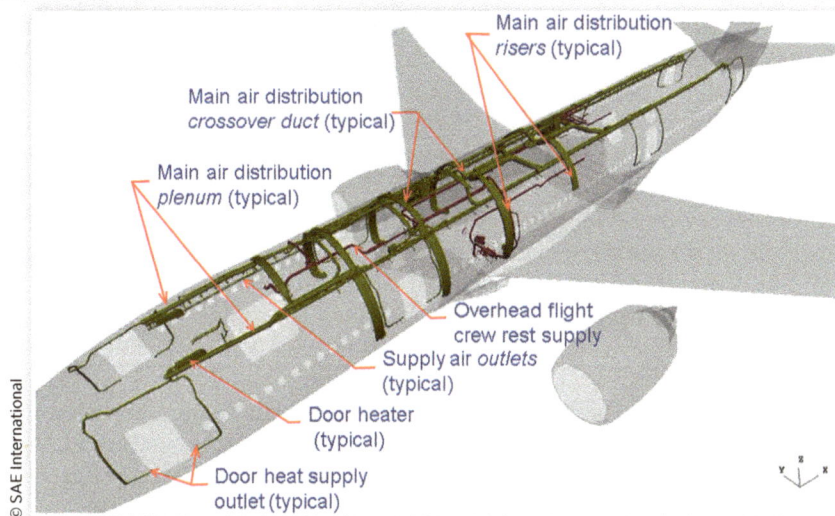

Air flows from the mix bay up the main air distribution *risers* through *crossover ducts* to the main air distribution *plenums*, which supply overhead air *outlets* (Figure 7.12).

FIGURE 7.12 Cabin airflow path.

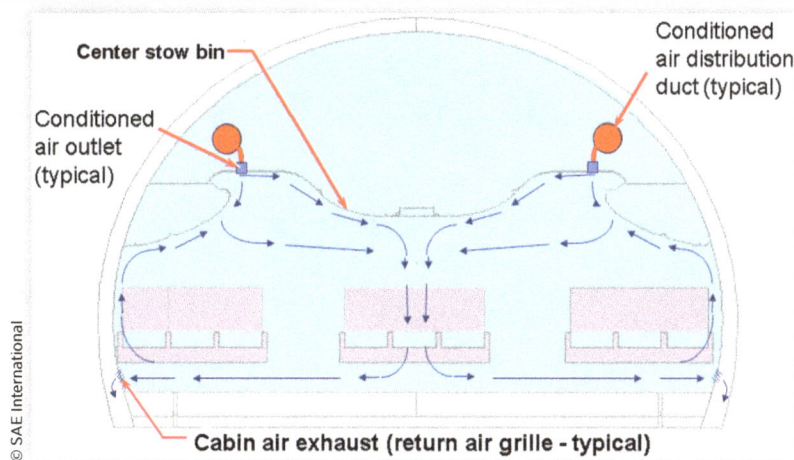

An effective air outlet design increases the amount of cabin air entrained by the supply air jet, ensuring greater mixing and more uniform air temperatures.

Negative pressure generated in the lower lobe, by recirculation fans and outflow values exhausting air from the lower lobe, drives flow downward through the return air grills. A well-designed air outlet will also send most of the air exhausted by each individual passenger toward the floor before passengers in adjacent seats have an opportunity to inhale it.

Air reaching the cabin exhaust grills is either cleaned by the recirculation system filters or exhausted overboard.

An airplane can have multiple passenger ventilation zones providing different supply temperatures based on the cooling and heating needs. A business class section with less dense seating usually requires less cooling, and a different supply air temperature, compared to the zones of densely packed (cheap) seats further aft.

Hot trim air added to conditioned air flowing from the mix bay accommodate these different supply temperature requirements. Some airplanes include individual air outlets to allow passengers to control the velocity and direction of high-velocity air. The air can come from fans blowing filtered cabin air or the air distribution system supply.

The main air distribution system also supplies *overhead crew rest* areas, an option for larger airplanes, and *door heaters*. Door areas can get colder than the seated passenger zones as additional supporting structure conducts heat to the external airplane surfaces. Electric resistance heaters mounted on the floor and/or a stream of heated supply air running from the air distribution ducting system provides door heat. Flight attendant fold-down seats are often located adjacent to doors making it particularly important to control these structure temperatures impacting cabin crew comfort.

7.2.3.3 FLIGHT DECK

The flight deck air distribution system (Figure 7.13) provides much more air per occupant than the passenger cabin to meet additional cooling needs. Heat from electronics and the sun beating though larger window areas oriented more perpendicular to the incident sun rays far exceed the heat generated in the passenger cabin for the same volume or occupant loading.

Mufflers can reduce noise generated by air flowing through ducts. This a major concern in the flight deck where passenger safety can depend on clear communication between crew members and flight controllers.

FIGURE 7.13   Flight deck air distribution.

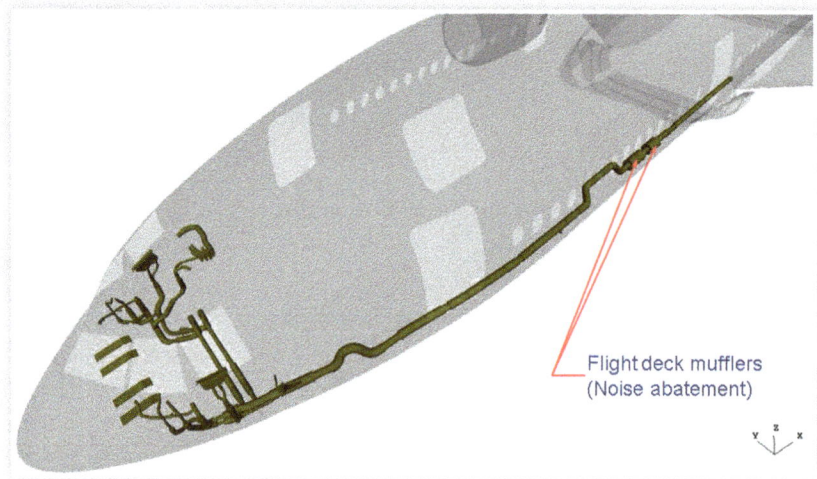

Flight deck mufflers
(Noise abatement)

© SAE International

7.2.4 Cargo Heat and Cargo Air Conditioning

A cargo heat system delivers the *electrical/electronic* (EE) cooling system's exhaust heat (air) or hot bleed air under the cargo floor to prevent subfreezing temperatures during flight. This protects cargo placed near the floor from freeze damage and prevents a slip hazard for baggage handlers due to water on the floor freezing during flight.

Safe transport of perishables and live animals may require the additional control (beyond staying above freezing) provided by a cargo AC system to ensure acceptable air temperatures, relative humidity, and *carbon dioxide* (CO_2) levels. The simplest cargo heat systems lack thermostats to control the amount of heat supplied to the compartment, which leads to wide compartment temperature swings.

A *cargo compartment AC system* delivers conditioned air and circulates the compartment air to accommodate the safe transport of animals, produce, flowers, some pharmaceuticals, dry ice, and other perishables. It provides a selectable temperature with cooling capabilities to offset heat generation from animals.

A cargo compartment AC system can be as simple as a duct providing conditioned air from the cabin air distribution system, or as complex as a dedicated VCM. Figure 7.14 shows a typical *forward cargo AC* system combined with a cargo heat system.

FIGURE 7.14   Forward cargo AC system.

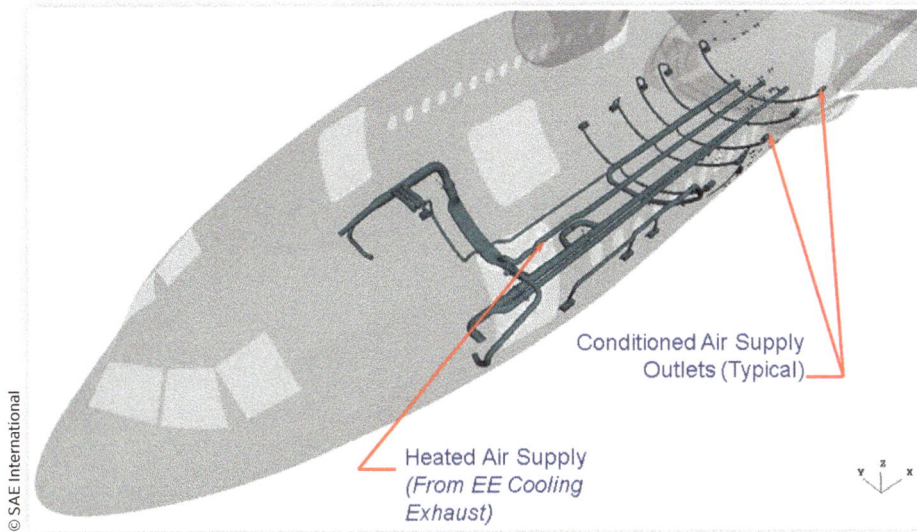

© SAE International

Conditioned Air Supply
Outlets (Typical)

Heated Air Supply
*(From EE Cooling
Exhaust)*

7.2.5 Cabin Pressure Control

Exhaust flow rates through one or two OFVs control the cabin pressure. During climb and initial cruise, the cabin pressure control system varies the rate at which cabin exhaust air exits the OFV(s) to ensure a gradual decrease in cabin pressure until reaching the minimum cabin pressure for the selected cruise altitude. This change in cabin altitude would cause some passengers" ears to pop prior to the use of today's precise electronic controllers and pressurization schedules optimized for passenger comfort.

The maximum allowable cabin altitude under normal conditions is 8000 ft (2438 m) per FAA requirements, although airplane manufacturers have designed aircraft for lower altitudes to improve passenger comfort (Section 1.5.1.1).

7.3 Venting and Chiller Exhaust

Venting systems transfer hot or odor-laden air overboard through airplane skin penetration or by blowing the exhaust at a lower lobe OFV (Figure 7.15).

A *lavatory* (lav)/galley fan (1A) exhausts air near an OFV (1A), which it draws from vents (2A) in the lavatories, galleys, and chiller exhaust (2). The airflow prevents odors

FIGURE 7.15 Airplane exhaust systems.

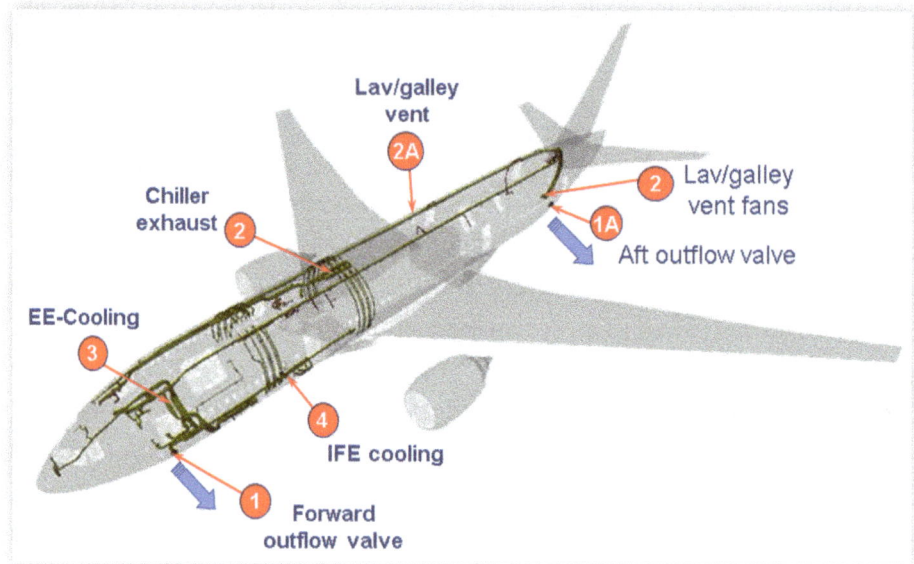

Lav/galley
vent
2A

Lav/galley
2 vent fans

Chiller
exhaust **2**

1A

Aft outflow valve

EE-Cooling
3

4

IFE cooling

1

Forward
outflow valve

© SAE International

generated in the lavatories and galleys from entering occupied areas on the main deck and prevents the chiller heat load from entering the cabin. Chillers supply food carts cool air.

7.4 **EE Cooling**

One or more lower lobe compartments, called EE bays (Figure 7.16), contain most of the EE equipment and associated cooling systems on large commercial airplanes.

FIGURE 7.16 EE bay.

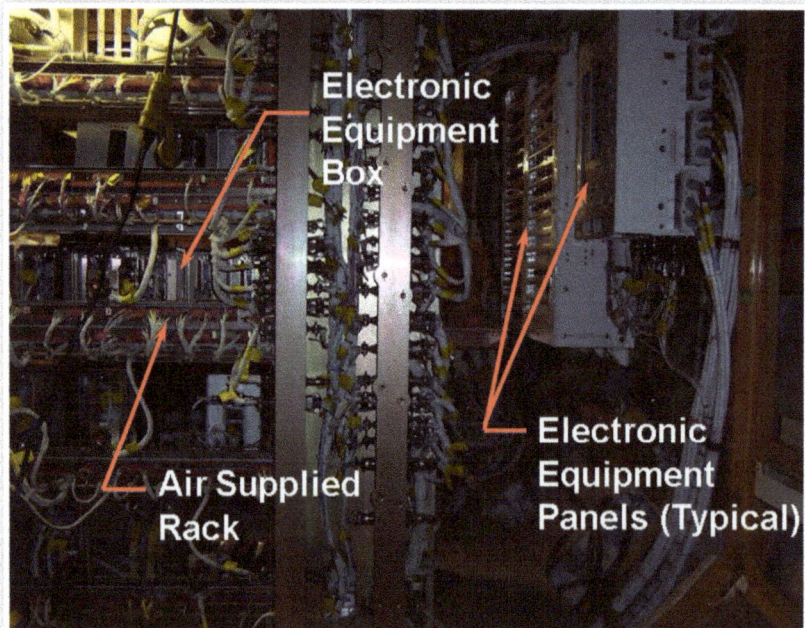

Electronic
Equipment
Box

Electronic
Equipment
Panels (Typical)

Air Supplied
Rack

© Boeing

Equipment is cooled using a combination of active systems and designs that enhance passive heat transfer from component surfaces.

7.4.1 Active Cooling

Active EE cooling systems pump a fluid through boxes enclosing the electronics to remove the heat for transfer to a heat sink. Cabin exhaust air is the preferred fluid since it eliminates concerns over fluid leaks and reliability associated with liquid cooling systems.

The wider allowable operating temperature range for electronics, compared to people, make it possible to reuse air that has cooled the passenger cabin to provide electronic cooling. Figure 7.17 shows the component locations and ducting for a typical forced-air EE cooling system.

FIGURE 7.17   Forced-air EE cooling system.

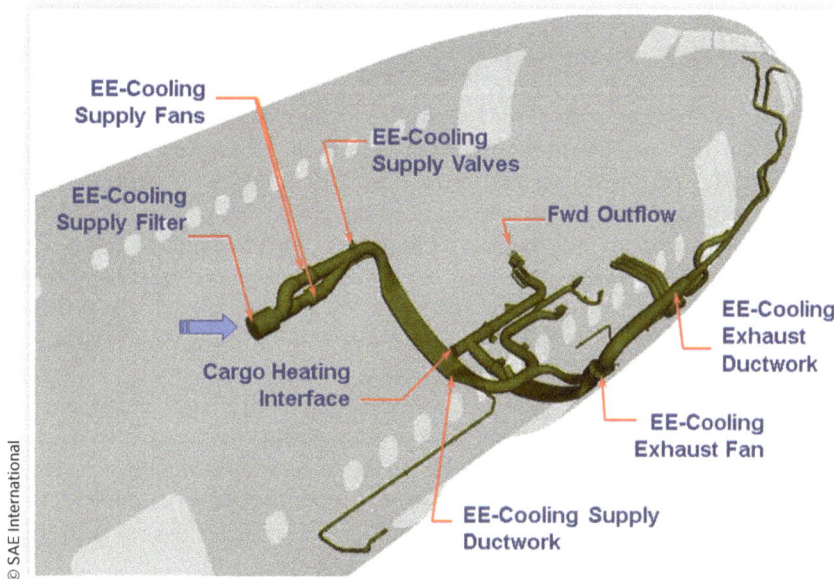

A forced-air EE cooling system includes both an *EE-cooling supply fan*, which blows air through electronic boxes mounted on rack shelves, and an *EE-cooling exhaust fan*, which pulls air through electronic boxes mounted on different shelves. The supply fan and attached ducting make up a *blow-through system*, while the exhaust fan and associated ducting create a *draw-through system*.

7.4.2 Passive Cooling

Passive cooling uses convective and radiative heat transfer from an electronic box's outer surface to transfer heat to the surrounding ambient. Some passively cooled racks have inner flow channels that allow buoyancy-driven air to remove heat. Air heated by the electronics rises, passively drawing in cooler replacement air.

Passive cooling is an appealing approach when practical since it eliminates the cost, weight, space, and power required for an active cooling system. Although, equipment designed for passive cooling is usually heavier due to the need for larger surface areas (fins) to remove the heat.

Since the EE bay removes revenue-generating cargo volume, there is a strong incentive to minimize its size. Therefore, passively cooled equipment may appear outside the EE bay in non-revenue-generating areas. Above the passenger cabin in the crown, or in the cheeks adjacent to the cargo compartments, are two common locations for passive equipment. Airplane system configurators will even place electronics in unpressurized areas. Reliability issues, however, may follow when equipment designed for a controlled environment of the pressurized fuselage faces more extreme temperatures in unpressurized areas. Airplane designers place some equipment outside the EE bay to reduce the length and weight of wire runs. As is often the case, it is up to the thermal team to identify a low airplane impact means for accommodating these designs, or adequate data to convince others to move equipment to more benign temperature locations.

While there was a push toward passively cooled electronics in the past, increasing power densities make it impractical for some of the latest electronics and electrical equipment. Coming up with more reliable liquid cooling systems, therefore, is the latest focus due to increased power densities accompanying electronic miniaturization.

7.4.3 Flight Critical Equipment

Flight critical equipment required for *continued safe flight and landing* (CSFL), which needs active cooling, may include connections to multiple EE cooling systems to meet stringent reliability requirements. One example is an electronic box connected to both a blow-through and draw-through system. Both the blow-through and draw-through systems include redundant fans so three of four fans could fail and the flight critical equipment would still receive cooling air in this configuration. If additional backup cooling is needed, a duct connected to the airplane's skin can provide draw-through cooling with the opening of an *overboard exhaust valve* (OEV) during flight. The pressure differential between the pressurized cabin and the outside ambient air can generate significant flow at a cruise altitude. If the OEV failed (to open), along with the four fans, flight critical equipment may still survive long enough for the airplane to reach the closest useable airport since the equipment must also meet *loss-of-cooling* requirements with no cooling airflow. This layer upon layer of redundant systems contributes to the remarkably high passenger safety record for commercial jet travel (Section 6.1).

The blow-through and draw-through systems also serve equipment located in the flight deck.

Liquid cooling is an option when equipment power densities are too high to remove sufficient heat with convective heat transfer to air, or the passenger cabin exhaust provides an insufficient heat sink. Liquid cooling systems transfer electronic-generated heat to a heat sink, such as fuel or outside ambient air (Chapter 5), by pumping a heat transfer liquid in a closed loop system consisting of a pump, tubing, and a compact HX. The compact HX is normally located in a ram-air system (Section 5.2.2.1) or fuel tank (Section 5.4.1).

7.5 Protective Systems

Excessive atmospheric ice buildup on external aircraft surfaces, such as wings, propellers, rotor blades, or engine intakes, can degrade airplane performance, jeopardizing safety. Ice on the wing can decrease lift and increase drag by changing its shape and increasing weight. Dislodged ice can damage airplane structure or system parts, such as spinning engine blades, which may reduce the thrust or even disable an engine. Ice blocking an environmental control intake, like pitot probes measuring airplane speed, can cause the pilot to fly below the speed required to maintain flight (stall speed). Each of these scenarios can quickly cause a crash. Meanwhile, ice or fog on a flight deck window can impede the flight crew's vision, restricting their ability to identify and react to hazardous conditions.

Ice can accumulate on an airplane when liquid strikes a surface below the 32°F (0°C) freezing point of water. This can occur on ground during a freezing rain or in flight when an airplane flies through clouds containing supercooled water droplets. Atmospheric water can exist as a liquid at a temperature as low as −4°F (−20°C) when the air temperature is between 32°F (0°C) and −40°F (−40°C) [6]. Consequently, certain atmospheric conditions can cause extensive ice accumulation.

Water in clouds consists almost entirely of ice crystals below an air temperature of −40°F (−40°C), so ice buildup on aircraft occurs largely during operations at lower altitudes during takeoff, climb, descent, or landing. Holding patterns, during delayed landings, can also contribute to extensive time in icing conditions, although pilots will normally change altitudes to minimize this hazard.

Prior to takeoff, the pilot can request deicing of airplane surfaces from trucks that spray a heated deicing fluid to remove ice and snow. Deicing fluids contain glycol and water. Spraying anti-icing fluids containing a higher concentration of glycol on aircraft surfaces provides additional protection for atmospheric conditions especially conducive to icing.

While the anti-icing fluid can reduce the ice buildup during the initial climb phase, it does not necessarily eliminate it for the remaining portion of a mission, driving the need for protective systems to either prevent (anti-ice) or remove (deice) airplane ice. This usually involves heat generation (on jet aircraft), making them an important part of aircraft thermal management.

Figure 7.18 lists airplane structure and systems served by ice and rain protection systems.

FIGURE 7.18　Ice and rain protection systems [7].

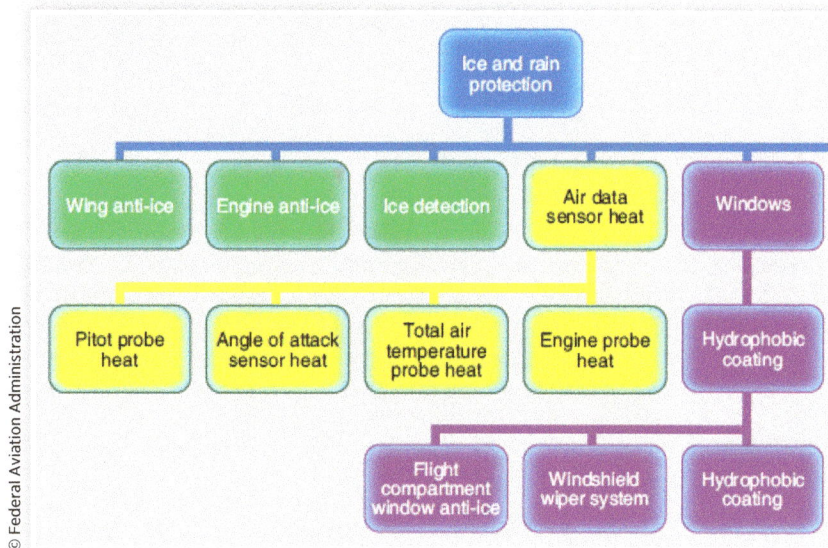

© Federal Aviation Administration

7.5.1 Wing Anti-ice

The wing leading edge of many aircraft have *wing anti-ice* systems to prevent ice formation. Jet-powered aircraft usually use heat (thermal energy) to either evaporate impinging water or raise the surface temperature just enough to prevent freezing.

7.5.1.1 THERMAL

The heat for ice protection typically comes from the engine bleed air or for no-bleed aircraft, like the Boeing 787, electric resistance heaters layered on top of or built into the leading edge surface. A holed (piccolo) duct that directs air on to the outer airplane's skin before venting overboard provides heated anti-icing or deicing air for most aircraft with bleed-air systems (Figure 7.19).

Thermal anti-icing and deicing require huge amounts of thermal energy or electrical power (usually for short time periods), leading to the development of less energy-intensive options based on chemicals or mechanical means.

FIGURE 7.19   Wing leading edge heated with bleed air [8].

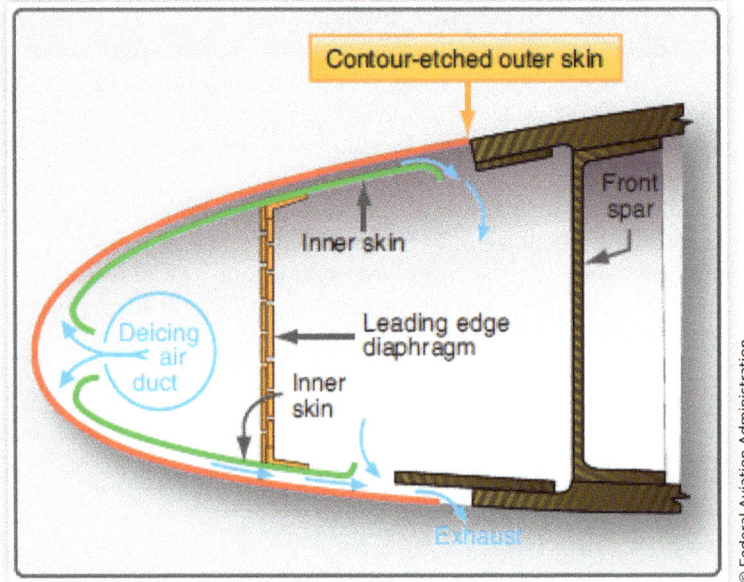

7.5.1.2 CHEMICAL

Some smaller aircraft use chemical anti-ice systems that pump environmentally friendly anti-freeze solutions through tiny holes in the leading edge creating a *weeping wing*. This solution, which has a much lower freezing point than water, mixes with supercooled water in the cloud impinging on the aircraft surface causing it to flow off rather than freeze. After ice has formed on the airplane surface, the anti-freeze solution chemically breaks down the bond between the ice and airframe, allowing aerodynamic forces to shed the ice to provide deicing protection.

7.5.1.3 MECHANICAL

Smaller general aviation aircraft often use inflatable rubber "boots" running along the leading surface that expand when inflated to knock ice off the wing surface. After removing the ice, suction applied to the boots returns them to their original shape for normal flight (Figure 7.20).

A second mechanical means for deicing a metallic airplane surface uses electromagnetic coils under the skin to rapidly accelerate the surface, which shakes the ice off. Hitting the surface with

FIGURE 7.20   Inflatable wing leading edge deicing boot [9].

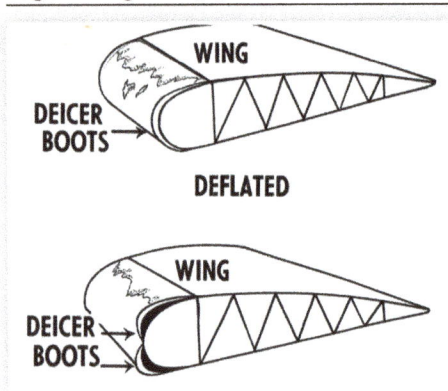

a hammer produces a similar effect. This method, called *electro-impulse deicing* (EIDI), induces strong eddy currents to move the airplane's skin.

While the chemical and mechanical ice protection, use much less energy compared to thermal systems, a few concerns may have prevented their penetration into the large commercial transport marketplace. The chemical system adds to the cost of the anti-icing fluid and a maintenance procedure to top off airplane fluid reservoirs. Rubber also degrades over time, adding another maintenance cost, and would be difficult to incorporate into the wing contours of a large commercial airplane. EIDI raises concerns over reducing the life of aircraft structure from the frequent flexing of metallic surfaces and will not work for composite surfaces.

Wing anti-icing systems also work on the leading edge of the vertical and horizontal stabilizers.

7.5.2 Engine Anti-ice

Ice accumulation on the engine-cowl leading edge (lip) can lead to ice ingestion in the engine from shedding. Turbofan engines also require smooth surfaces to support laminar flow, which maximizes engine performance. Consequently, engine cowls include anti-icing systems which operate like the wing thermal anti-ice by either blowing hot air or using electric resistance heating.

7.5.3 Ice Detection

Some civilian transport aircraft offer ice detectors to warn the crew of icing conditions or automatically turn on the anti-icing or deicing systems. The upper horizontal wings surface can also include narrow stripes painted black running close to the fuselage to provide a visual indication of ice.

7.5.4 Air Data Sensors

Airplanes include multiple sensors to indicate aircraft speed and orientation (Figure 7.21).

FIGURE 7.21 Probes with thermal electric anti-icing [10].

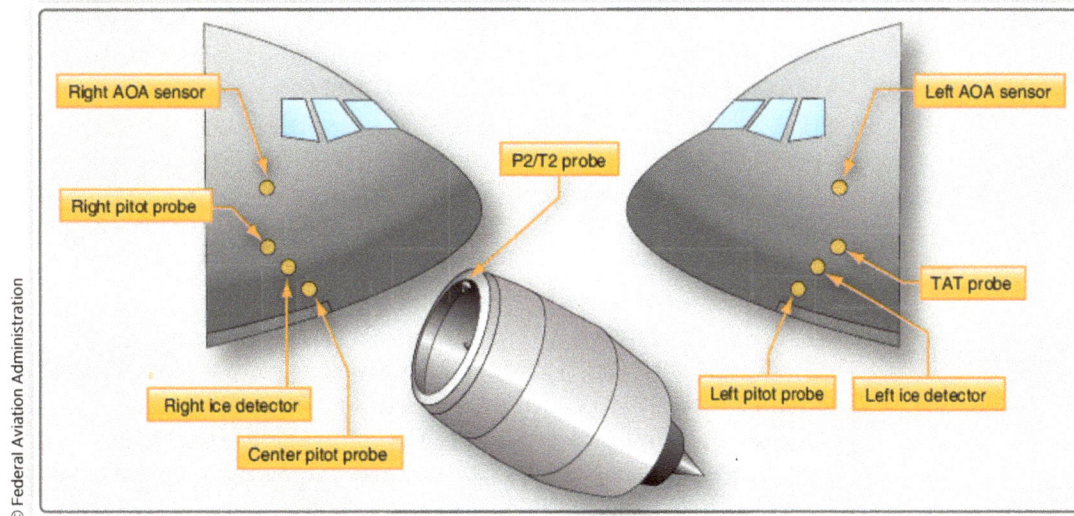

© Federal Aviation Administration

These probes are arguably the most critical instrumentation on an aircraft since faulty measurements can cause a pilot to inadvertently direct the airplane toward the ground leading to a crash. These sensors usually include extremely reliable electric resistance heaters designed to prevent ice formation under the most extreme cold conditions possible, from ground to the maximum cruise altitude. Some heaters automatically turn on at engine start regardless of the *outside ambient temperature* (OAT) or airplane movement, generating extremely high temperatures on adjacent airplane structure much of the time. To reduce the risk of failure, the same heat input designed to keep the sensor temperature above freezing, 32°F (0°C), while flying at Mach .86 in a −40°F (−40°C) TAT, may also occur during ground operations at Kuwait City, when the OAT is 120°F (49°C). Control logic to sense OAT and vary the heater power provides additional opportunities for failure that can produce unacceptably low reliability numbers.

7.5.5 Windshields

Large commercial airplanes have systems to keep flight deck windshield frost, fog, and ice free, using electric, pneumatics, and/or chemicals.

A typical flight windshield consists of a laminate of multiple layers of transparent plastic and glass designed to withstand the large forces of a pressurized cabin at altitude, or bird strike close to the ground. Pressure differences between the outside ambient air and the flight deck (and cabin) air can approach 9 psi (62 kPa) at the maximum cruise altitude for a commercial jet. Aircraft windshields must also survive the impact of a 4 lbm (1.8 kg) frozen chicken shot from a specially designed cannon, for certification.

The windshield design supports incorporation of a heating element or laminated conductive films within the glass layers. Some systems can operate at multiple power levels to better accommodate variation in outside ambient conditions. Blowing heated air on the windshield's interior surface is also a common design. Liquid window ice protection systems, which operate like the *weeping wing*, are less common.

Window heating systems must minimize temperature gradients occurring over the window surface (when operating) or risk generating excessive thermal stresses, a leading cause of cracks.

References

1. Wikipedia, "Bypass Ratio," 2017, accessed May 20, 2017, https://en.wikipedia.org/w/index.php?title=Bypass_ratio&oldid=816477985.

2. United States Department of Transportation/Federal Aviation Administration (USDOT/FAA), *Aviation Maintenance Technician Handbook: Airframe*, FAA-H-8083-31 (Oklahoma City, OK: Flight Standards Service, 2012), Volume 2, 16–35, ISBN 978-1-56027-953-2, https://www.faa.gov/regulations_policies/handbooks_manuals/aircraft/amt_airframe_handbook/.

3. USDOT/FAA, 2012, 16–45.

4. Wikipedia, "Vapor-Compression Refrigeration," 2017, accessed July 25, 2017, https://en.wikipedia.org/w/index.php?title=Vapor-compression_refrigeration&oldid=810107257.

5. Peterson, B., "Airplane Air: Not as Bad as You Think," *nbcnews*, January 13, 2010, http://www.nbcnews.com/id/34708785/ns/travel-travel_tips/t/airplane-air-not-bad-you-think/#.WM3hnW_yuM8.

6. Flight Safety Foundation, "Supercooled Water Droplets," *Skybrary*, Last modified July 30, 2017, accessed August 15, 2017, http://www.skybrary.aero/index.php/Supercooled_Water_Droplets.

7. USDOT/FAA, 2012, 15-3.

8. Ibid., 15-5.

9. Wikipedia, "Deicing Boot," 2017, accessed September 20, 2017, https://en.wikipedia.org/w/index.php?title=Deicing_boot&oldid=801503442.

10. USDOT/FAA, 2012, 15-11.

Thermal Design

"You can be like a thermometer, just reflecting the world around you, or you can be a thermostat, one of those people who sets the temperature."

—**Cory Booker [American politician]**

8.1 Introduction

Aircraft thermal design involves the installation of measures to minimize the extent and frequency of extreme temperatures, and variations faced by passengers, crew, cargo, equipment, and structure.

Extreme (or merely elevated) temperatures tend to reduce the performance, efficiency, and reliability, and increase the cost of virtually everything on an airplane, from the smallest processor in an electronic box to the entire structure. Developing *low-cost*, *lightweight*, efficient thermal designs, which moderate or accommodate high or low temperatures, is the prime priority of *aircraft thermal management* (ATM) in the eyes of aircraft engineers responsible for systems and structure impacted by temperatures. This covers just about everyone.

Controlling heat generated within the airplane (Chapter 3) and externally (Chapter 4) is the primary means for maintaining more moderate aircraft structure, system, and pressurized volume ambient temperatures. Thermal design options include *limiting the transfer* of, *removing*, *reducing the intensity* (spreading it over a larger area) of, *controlling the rate of release* of, or *changing heat to a different form*.

8.2 Insulation Types

Insulation is the most widely used design for limiting heat transfer on an aircraft because it is normally one of the cheapest, lightest, and least technically risky options. It also reduces noise transmittance, which can be as important as reducing heat transfer on commercial aircraft. Consequently, the term *thermoacoustics* often describes aircraft insulation.

Commercial airplanes use insulation on the fuselage skin, bleed-system ducts in the leading edge, hot ducts and components in the pack bay, air distribution ducting inside the fuselage, and occasionally structure adjacent to heat-generating components.

Thermal conductivity (k), density (ρ), and operating temperature ranges vary widely between materials used on an aircraft (Figure 8.1), due to the highly variable nature of thermal protection needs.

FIGURE 8.1   Insulation material thermal conductivities. Source: [1, 2, 3, 4, 5, 6].

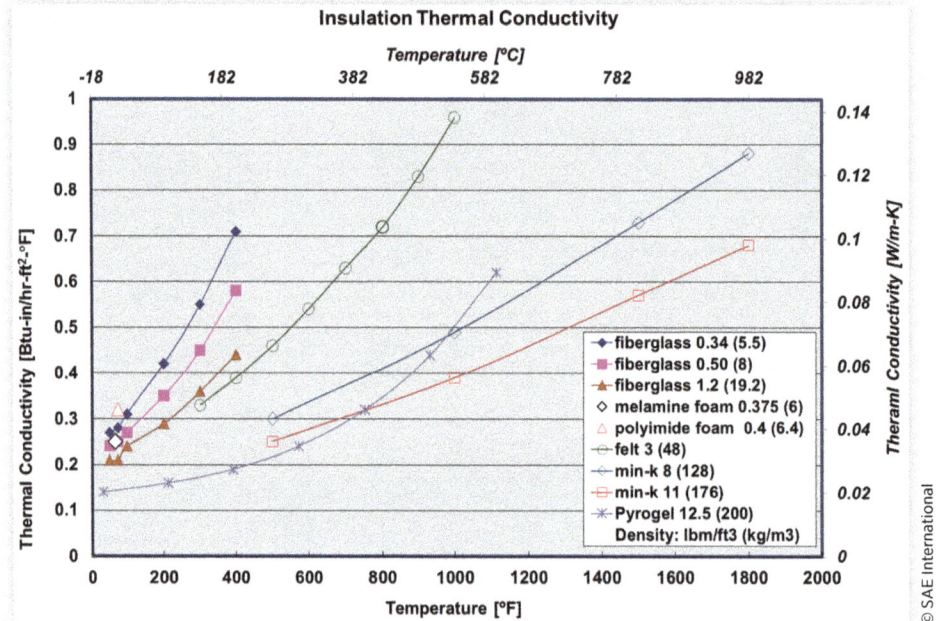

© SAE International

8.2.1 **Fiberglass**

Glass fibers coated with a resin and cut to a length before placement in bag, form fiberglass blankets. This is the same material used to insulate most homes, after modifications to reduce weight and improve performance. It is the most commonly used insulation material on commercial airplanes due to extremely low-cost, low k relative to weight, and decent noise attenuation performance. The biggest difference between the fiberglass blankets sold at the big box stores and those installed on aircraft are the added covering materials and ρ. A fiberglass blanket's outer cover (bag) material contains the fibers, *prevents liquid water ingression*, and *reduces radiative heat transfer during a fire*.

Since glass fibers can absorb water, the containment bag must be impervious to liquids while allowing water vapor to escape, similar to the Gor Tex® "breathable rain gear." Liquid water can enter insulation bags through poorly sealed seams or tears from careless installations. While most of the liquid ends up leaving through drainage holes located at the lowest point on the bag, the glass fibers may absorb some water. Even fibers with hydrophobic coatings designed to repel liquids can absorb water as age and dirt degrades their effectiveness, leading to *wet blankets*. Wet blankets are a major concern for airlines due to the added weight and the risk of water dripping onto occupants in the cabin (for crown insulation).

Water absorbed by the fibers can only evaporate (drying out the blankets) if the blanket covering material allows vapor transport to the lower humidity cabin air flowing against the bag surface. Moisture in air flows from locations of higher to lower concentrations.

This ability to absorb liquids can create a safety issue when it involves flammable liquids close to an ignition source, like an insulated bleed (hot) air duct located underneath a leaking *center wing* fuel *tank* (CWT). Consequently, fiberglass may be unacceptable for *flammable leakage zones*, which prohibit artifacts capable of collecting liquid. The covering material must also limit radiant heat transfer during a fire to meet certification requirements.

While fiberglass is cheap and has impressive thermal resistance numbers relative to its weight, it has one other disadvantage beyond a tendency to absorb liquids. The low ρ, 0.34 lbm/ft^3 (5.5 kg/m^3) premium brand is a delicate material that loses loft under the smallest force. Therefore, a 1" thick fiberglass blanket could significantly compress after installation. Especially when wrapped on curved surfaces of a smaller diameter duct.

Higher ρ (heavier) fiberglass improves loft retention (thermal performance) and noise attenuation. While the higher ρ insulation materials also have reduced k-values, the effect is small. Therefore, spending additional weight on a thicker blanket is a more efficient way for reducing heat transfer.

8.2.2 Open Cell Foams

Open-cell foam insulations that exhibit similar thermal resistance per lbm as fiberglass reduce the problem of poor loft retention since the lightest weight versions compress relatively little. They also require covering materials that meet the same requirements as fiberglass blanket bags. Open-cell foams absorb liquids and are vulnerable to destruction from a fire similar to fiberglass. The encasement method, however, differs with a very thin covering material fused on both the inner and outer faces of the foam sheet.

Two common lightweight Open-cell foam insulations used for aircraft installations are polyimide and melamine.

8.2.2.1 POLYIMIDE

Polyimide is a polymer of imide monomers used to produce lightweight, flexible materials resistant to heat and chemicals. Developed for spacecraft by the United States (US) *National Aeronautics and Space Administration* (NASA), polyimide foam meets stringent performance requirements similar to aircraft applications.

Typical uses of polyimide foam are insulating ducts or aircraft structure. The lower ρ versions provide slightly higher published k-values compared to the equivalent weight fiberglass, which means little after accounting for loft retention. The lightest weight polyimide foam compresses much less after installation (compared to fiberglass), providing a more certain and higher thermal resistance in service. The open cells, which allow liquid ingress, may also prevent its use in flammable leakage zones.

The main disadvantage of polyimide foam is its high cost, in addition to a tendency to absorb liquids. High costs led to a search for cheaper materials providing similar performance, such as melamine foam.

8.2.2.2 MELAMINE

Melamine foam consists of a formaldehyde-melamine-sodium bisulfite copolymer material that provides the "magic," in the "magic erasers," used by millions to remove wall markings.

Melamine foam offers similar or slightly better thermal performance relative to weight and loft retention (of polyimide foam) at a lower cost. Competition among numerous manufacturers keep melamine prices lower than polyimide but higher than fiberglass.

8.2.3 Closed Cell Foams

Closed-cell foam insulations perform much better than open-cell foams in every area but weight, and possibly costs. The closed cells limit moisture absorption, making them well suited for flammable leakage zones, which have fewer insulation options. Lower k-values provide an advantage for space-constrained installations, which frequently occur on aircraft. They are also more durable (better able to bounce back from baggage handler footsteps) then fiberglass and open-cell foams.

8.2.4 Ceramics

Ceramic fiber and microporous (Min-K®) insulation provide thermal protection at much higher temperatures then the lighter weight alternatives, with maximum continuous operating temperatures exceeding 1800°F (982°C). This makes them suitable around engine compartments, where insulation must be able to survive a raging fire. Ceramic tiles protected portions of the US Space Shuttle orbiter structure from re-entry heating.

The main disadvantage of ceramics is a much higher weight, a common problem with higher temperature insulation materials.

8.2.5 Felt

Felt made from silica fibers can provide continuous thermal protection up to 1800°F (982°C), using an unusually low-ρ material, at 3 lbm/ft³ (48 kg/m³), for such high temperatures. The k of felt is, however, greater than for the denser Min-K® ceramics used for similar applications.

8.2.6 Aerogels

Aerogels are a class of materials famous for being the world's lightest solids. Silica aerogels exhibit the lowest k of any known solid. Extreme fragility has limited dreams of creating amazingly light and low k insulation materials from aerogel.

Heavy containment mediums solved the durability problem, so aerogel now provides superior thermal performance (relative to thickness) in clothing and insulation systems. Unfortunately, much of the hoped-for weight savings are sacrificed in the containment mediums.

Aerogel products providing impressively low k are heavy, limiting their use to small areas with extreme space constraints. For example, the k of a commercially available aerospace aerogel (Pyrogel XTE) is 0.194 Btu-in/h-ft²-°F (0.028 W/m-k) at 392°F (200°C), about one-fourth the value of the lowest ρ fiberglass (Figure 8.1). The Pyrogel ρ, however, is almost 40 times greater, 12.5 versus 0.34 lbm/ft³ (200 versus 5.5 kg/m³). If the fiberglass compresses to half its installed thickness, that still leaves a five-fold weight penalty to provide an equivalent thermal resistance using the more expensive aerogel.

8.3 **Insulation Applications**

8.3.1 **Fuselage**

Insulation covering the fuselage's internal skin surfaces limits heat transfer between the conditioned air and outside ambient air and maintains sidewall surface temperatures close enough to the cabin ambient air to ensure passenger comfort. Insulation also attenuates noise reaching the cabin from the engines, and air flowing over the external surfaces on commercial airplanes. Since a quiet cabin is an important passenger aircraft marketing tool, noise attenuation needs may drive an insulation thickness beyond the thermal requirements, or the use of a heavier material. Mass tends to reduce the transfer of noise.

Fuselage insulation also provides other benefits beyond reducing heat and noise transfer. When held in contact with the skin, it can reduce airflow, cutting down on the opportunity for moisture to condense on the colder surfaces during flight. Condensation on metal structure can accelerate corrosion; drip on passengers or crew in the cabin, electrical/electronic boxes, or power panels in the lower lobe; or infiltrate insulation blankets (adding weight and reducing thermal resistance).

The advantage of reduced moisture condensation on the skin during flight is traded against the negative effect of reduced airflow to dry out the skin and blanket during certain mission times. Cabin air is often very dry, providing ideal conditions for drying out insulation blankets if an open path is available for airflow.

Fuselage insulation must also provide adequate thermal protection (and prevent flame penetration in combination with the airplane's skin) from an external fire, to provide passengers and crew enough time to escape a survivable crash and subsequent fuel fire outside the aircraft. The 0.1-in. (0.254 cm) thick aluminum skin, which protects passengers from low ambient pressures during a high-altitude cruise, melts between 1025°F and 1200°F (550 and 650°C). This is no match for the 1800°F (980°C) flames of a raging fuel fire. Insulation (and the covering material), therefore, may provide the additional time required to allow all passengers and crew to escape a fuel fire.

Ironically, composite structure, which loses strength at much lower temperatures than aluminum, provides more thermal protection during an external fuel fire. Long after aluminum structure melts away, the composite fibers are still in place preventing flame penetration. Although, there are the potential negative consequences for breathing gases entering the cabin as the resins holding the carbon fibers together burn up.

The insulation must also resist mold, even after an extensive time spent saturated with water, meet stringent toxicity requirements, and avoid falling apart after years of engine-induced airplane vibrations.

8.3.2 **Ducting and Hot Pack Components**

Insulation covering hot-air ducts, such as bleed air and some pack bay ducts, and hot components (i.e., HXs, compressor motors and housings, ozone converters, valves) prevents excessive heat transfer to adjacent structure, systems, or air. Additional protection (beyond insulation) may be needed to shield adjacent surfaces from impinging high-pressure and -temperature air during a duct or component failure.

Air temperatures above 300°F (149°C) usually accompany higher pressures with destructive power when unleashed. While open-cell insulations designed to operate at these temperatures are available, they provide little protection following a duct burst (i.e., weldment failure, coupling leakage, or large crack). The explosive impact of a rupturing duct may require a much denser insulation, with a metallic covering material for containment. Afterward, the continuous wire or eutectic loop used for the main landing gear fire detection system (Section 6.2.1) can detect hot air filling the volume from an elevated air temperature.

Enclosing a duct with a second duct provides more protection from the energy released from a burst duct, compared to an insulation blanket. It may, however, provide less thermal protection during normal operations, involve a more complex installation, and weigh more.

Air distribution ducting insulation reduces heat gain from the pressurized cabin air, on its way from the *air-conditioning* (AC) packs to the cabin overhead air distribution outlets. Design requirements are less demanding than for high-temperature duct insulation because the lower air pressure and temperatures provide no threat to structure during a failure. The insulation also reduces noise generated by air flowing through the ducting as does limiting the air velocity.

8.3.3 Engine and Auxiliary Power Unit (APU)

The engines and APU generate the highest temperatures affecting structure on an aircraft. Fire detection and suppression systems and very robust insulation mitigate the fire risk. Ceramic insulation enclosed in a metallic covering material to limit radiant heat transfer and prevent flame impingement during a fire is a common thermal design.

8.3.4 Cargo Compartments

Cargo liners must contain flames and gases from a cargo fire for US *Federal Aviation Administration* (FAA) certification. This differs from fire thermal protection, since there are no requirements to limit heat transfer from cargo fires to surrounding structure imposed by the FAA. There are, however, structure temperature requirements imposed by the aircraft manufacturer, to meet the FAA requirement for *continued safe flight and landing*.

The cargo liner, a flexible fiberglass material, is too thin to provide significant thermal protection beyond limiting flame penetration.

8.3.5 Insulation Placement: Heat Source or Receiver

The most efficient and effective location for insulation placement depends on the situation. Locating insulation on a heat source (rather than the receiving surface) provides the added advantage of reducing the convective and radiative heat load to the compartment. Reducing the component heat dissipation may, however, reduce component life, by raising its temperature, creating a new problem. The insulation could even invalidate component certification testing, adding a large new cost for recertification.

8.4 Surface Coatings and Applications

Specially formulated coatings or material selection can limit heat transfer from solar and infrared radiation, when insulation is impractical, less efficient, or has a more negative impact on the airplane.

8.4.1 Low Solar Absorptivity Paints

Insulating airplane external surfaces is impractical, leaving the surface *solar absorptivity* (α) as the only way to limit solar-driven structural temperature increases.

Darker colored paints and surfaces absorb more solar energy than lighter ones, increasing structure design temperatures on aircraft sitting on ground prior to takeoff. Limiting aircraft external paint choices to a lighter color, ideally white, is an effective technical means for minimizing solar-driven structure peak temperatures, but a very poor marketing strategy. The building industry faces a similar issue dealing with the impact of dark roofs on cooling needs.

Roofs with high α surfaces may increase building AC loads. This led to the development of low α (high reflective) *cool* paints, which are available from multiple manufacturers.

The sun emits energy at wavelengths in the ultraviolet, visible, and infrared spectra. Only half of the energy occurs in the visible range where the wavelengths determine the color your eyes perceive (Section 4.2.1). This means a paint tailored to absorb less of the remaining 50% of the energy would not affect the appearance. Researchers have used that knowledge to formulate paints with impressive reductions in α, such as the Table 8.1 roofing tiles.

TABLE 8.1 Low solar absorptivity (high reflective) paint performance [7]

	Solar reflection	Solar absorption	Solar heat grain	Percentage
Graphite	11%	89%	801 W/m²	100%
Cool Graphite	43%	57%	513 W/m²	64%
Grey	17%	83%	747 W/m²	100%
Cool Grey	47%	53%	477 W/m²	64%
Beige	45%	55%	495 W/m²	100%
Cool Beige	61%	39%	351 W/m²	70%

© SAE International

While solar heat-reflective aircraft coatings appeared at the Paris Air Show back in 2006 [8], they have yet to see widespread application due to unsolved issues.

Adoption of *low solar load* paints, to reduce structure temperatures and the associated weight penalty, includes the risk of in-service degradation and costs. Structure dependent on a specially formulated paint for adequate strength to meet certification requirements adds the economic burden (to the airlines) of maintaining a special paint and paint inspection procedures. This becomes a greater concern when a fledgling airline in the developing world, which may lack the rigorous oversight to ensure that proper maintenance procedures continue, purchases the airplane.

8.4.2 Low Emissivity Coatings

Radiation is usually the primary means a hot duct transfers heat to adjacent structure or systems. An extremely thin layer of a low *emissivity* (ε) material (such as gold) plated on a hot metal duct, therefore, may provide as much (or more) thermal protection as insulation in areas with limited space. Surface coatings also weigh less, eliminate the installation time for adding insulation, and prevent the potential problem of insulation coming loose in service. Loose insulation is a major concern for ducts running through the main landing gear wheel well with open doors generating a 125 mph (200 km/h) or more wind (the airplane speed) during takeoff and landing.

Gold-coated ducts, however, are very expensive and their thermal performance degrades over time as contaminants accumulate.

8.5 **Radiation Shields**

Shields with a low ε surface are another effective means to block radiative heat transfer. They usually consist of a metallic sheet, normally aluminum, which has a naturally low ε value. Often used between engines and surrounding structure, on both military and civilian aircraft, they also prevent the plume generated from the heat source from impinging on surfaces directly overhead.

They suffer the same disadvantage of reduced effectiveness following contamination as the gold-coated duct, in addition to weighing much more, and providing installation challenges.

8.6 **Phase-Change Materials**

Phase-change materials (PCMs) absorb and release heat at a constant temperature in changing between liquid and solid phases. Water provides a phase-change example observed by many. In freezing, water initially at 32°F (0°C) releases as much (latent) heat as the same amount of liquid dropping 144°C (80°C). Some farmers spray their crops with water on nights when freezing temperatures are predicted, to heat the vegetation from the latent energy released as the water freezes on the plants. The ice also provides a thermal capacitance which slows the temperature change of the encased plants.

Since heat transfer is proportional to temperature differences, a constant temperature heat sink provides more predictable and consistent heat transfer compared to a thermal mass whose temperature increases in time as it absorbs heat.

Satellites, which face extreme variations in the solar load in passing in and out of the earth's shadow, or perhaps a military weapon generating a massive but brief heat load, are likely candidates for PCMs.

8.7 **Intumescent Paints**

Intumescent paints expand (up to 100 times or more) upon reaching a specific temperature [9]. This process consumes some of the input energy in generating a thick char, which provides a thermal resistance to a heat source or flame. While intumescent paints are mostly used for fireproofing building structures, formulations which expand at much lower temperatures then a 1000°C (1832°F) flame, are also available. These coatings are only appropriate for protecting aircraft systems or structure from a failure condition.

8.8 **Ablation Materials**

Ablation materials vaporize to remove heat upon reaching extreme high temperatures. They are very reliable, extremely heavy, good for a single use, and a common thermal protection option for spacecraft re-entering the atmosphere with a payload. Astronauts and cosmonauts are the typical payload. Fire protection is the only logical use for ablation materials on an aircraft.

8.9 **Increase Heat Sink**

Increases in aircraft system heat loads, usually due to equipment upgrades, can drive the need for an increased heat sink (Chapter 5) to accept additional waste heat.

Aircraft system designers have three primary approaches for increasing their heat sinks. They can increase the cooling capacity of their existing ram-air systems (Section 5.2.2), improve the cooling capabilities of their fuel tanks (Section 5.4.1), or add a totally new system.

8.9.1 Ram Air

Letting the ram-air exhaust air temperature rise, or increasing the ram airflow, will increase the heat sink capacity of the ram-air circuit.

Ram-air system exhaust temperatures will naturally increase by adding more heat to the same flow stream. Higher ram exhaust temperatures will then increase the temperature of airplane structure affected by the exit plume and ram-air system ducting.

If higher exhaust temperatures are unacceptable, due to structural or equipment temperature limits, increasing the ram-air flow rate to absorb the additional waste heat, while maintaining the same exhaust temperature, is another option for increasing ram-air cooling. The on-ground fan pressure rise and in-flight ram-air pressure recovery must increase, or the ram-air system flow resistance must decrease to increase the cooling airflow rate. Neither is normally easy to accomplish. In a best-case scenario, merely adding more power increases the fan performance sufficiently to meet the increased flow requirement. "Merely" means also ensuring the availability of spare airplane-generated power; adequate current carrying capacity for wires supplying the fan; acceptable thermal impacts to adjacent air, systems, and structure; and an acceptable noise level for the higher fan speed. During flight, additional pressure may be possible by redesigning the inlet to capture more of the ram-pressure rise. This is only possible if the existing inlet is sufficiently under-designed.

Design changes that reduce the ram-air system flow resistance, to allow more airflow for the same pressure rise, are rarely practical.

8.9.2 Fuel

Raising the maximum allowable fuel temperature and increasing the ambient cooling rate are two options for increasing the fuel heat sink. Higher fuel temperatures, however, increase fuel flammability and reduce heat transfer rates from systems using the fuel as a heat sink (usually the engine oil cooler and hydraulic system).

8.9.2.1 FUEL FLAMMABILITY

The space above the liquid fuel in a tank contains a fuel vapor/oxygen mixture (ullage) that can ignite given an adequate *oxygen* (O_2) concentration and ignition (heat) source. Wires carrying power to a fuel quantity sensor or pump are potential ignition sources if their insulation wears away, or they come loose from a connector, allowing arcing or a spark to occur.

Design attempts to eliminate ignition sources in fuel tanks appear to have failed multiple times, most famously causing a fuel tank explosion and crash of a 747-100 (TWA Flight 800) shortly after takeoff in 1996 (Section 6.2.6).

The FAA's response to this tragedy was to require commercial transports with a passenger capacity of 30 or more to maintain the flammability risk of the worldwide fleet of an aircraft model below a specified value. This meant aircraft manufacturers had to either lower the maximum fuel temperature in their tanks or reduce the oxygen content to provide a reduced flammability risk. The acceptable risk is based on a probability of an aircraft flying a mission profile generating sufficiently high fuel temperatures in combination with a system failure causing a spark, with adequate energy to ignite the ullage.

8.9.2.2 NITROGEN-GENERATING SYSTEMS

Since lowering fuel temperatures for the existing in-service fleet of thousands of aircraft was unrealistic, the only alternative was reducing the O_2 levels in the fuel tanks by replacing some of the air with an inert gas. This new flammability requirement, which applies to airplanes certified since 1958, lead to the development of *nitrogen-generating systems* (NGS) for commercial airplanes. An NGS removes O_2 from outside ambient air to generate *nitrogen-enriched air* (NEA) to replace more flammable (higher O_2 concentration) ullage when distributed into a fuel tank.

Military aircraft use fuel tank inerting systems generating NEA with higher *nitrogen* (N_2) levels to offset the effects of higher fuel temperatures and ignition source energy levels, compared to commercial airplane systems. Fighter aircraft fuel tanks face higher heat loads relative to the amount of fuel carried, higher ambient boundary temperatures due to faster speeds (supersonic), and more frequent and severe ignition sources from incoming armaments.

8.9.2.3 SYSTEMS COOLING

Reductions in engine performance or life are possible following significant increases in the fuel temperatures due to reduced engine cooling. Meanwhile, hydraulic system fluids and seals can deteriorate from excessive time at elevated temperatures, due to inadequate heat transfer to the cooler fuel. These problems are primarily a concern for operations from extremely hot airports where warm fuel is loaded. Jet fuel is normally stored in aboveground tanks where the temperature tends to approach the daily average *outside ambient air temperature* (OAT).

8.9.2.4 INCREASING AMBIENT COOLING

Heat transfer from warmer fuel to the cooler outside ambient air, occurs primarily from convection to the upper and lower wing's wetted surfaces. Minimal heat transfer occurs from the fuel vapors (ullage) in the tank to the cold aircraft skin due to low convective heat transfer coefficients. Fuel consumption during a mission decreases the wetted wing surface area, reducing ambient cooling as the thermal capacitance for absorbing waste heat is also dropping. While the fuel thermal capacitance loss is unavoidable, pumping fuel over the dry surfaces of the uncovered lower wing can stem losses in ambient cooling. Military aircraft have used this approach for years to cool fuel in their wing tanks.

8.10 Reduce Heat Generation

Gordon Moore, cofounder of the computer chip manufacturer Intel, observed that the number of components per integrated circuit, which had been doubling every year back in 1965, would continue for the following decade. In 1975, he changed that to a prediction that the number of components would double every 2 years, which occurred for the following 40 years, leading to the famous *Moore's Law* [10].

Meanwhile, Jonathan Koomey, an energy specialist at Stanford University and Lawrence Berkeley National Laboratory, noticed that the number of computations per joule of energy spent has been doubling every 1.57 years since the 1950s, which is less well-known as Koomey's Law [11].

Slowing the growth in computing power used to a level below the natural increase in computing efficiency is one way to reduce the waste heat produced on an airplane.

A similar trend (at a much slower rate) is also occurring with aircraft power conditioning equipment, such as *direct current* (DC) to *alternating current* (AC) power converters, electrical generators, and motors. Small variations in the efficiency of a power converter carrying 250 kW can make a substantial difference on the amount of waste heat removal required. This is especially important for *more electric aircraft* (MEA) *architectures* (Section 1.5.3), which use multiples of the power for systems operation compared to a traditional aircraft.

8.11 Spot Cooling

Blowing air over a heat source or receiver is another means for temperature control. Possible air sources include the outside ambient air, the AC pack outlet, air distribution system, or cabin.

8.11.1 Unpressurized Air

The ram-air pressure rise used to supply ram-air cooling for the AC packs in flight, or a dedicated fan for ground operations, can supply ambient air for spot cooling in unpressurized areas. This approach, however, provides no ground cooling.

8.11.2 Pressurized Air

The AC pack outlet provides very cool air (at times), which takes much energy (fuel) to compress the outside ambient air and remove the heat generated in compressing the air. Using this conditioned air for spot cooling (beyond a tiny amount), before it serves the purpose of meeting the cabin ventilation and temperature conditioning requirements would, therefore, require unusual circumstances to justify.

The same penalties apply to using air distribution system air to a lesser degree. Air distribution air typically includes 50% recirculated air, which cuts the fuel burn penalty approximately in half.

A fan-blowing cabin air reduces the fuel burn penalty to the power required to run the fan and weight for the equipment and wiring. While running power to a new component is never a trivial task, wires are often easier to route than a duct supplying air.

8.12 Modify Material

Adding a higher *k* material to a lower *k* composite material layup can reduce the intensity of heat buildup from a point energy source. Composite wings follow this approach for lightning protection by incorporating an extremely thin outer metallic layer of *expanded metal foil* (EMF) [12].

The EMF spreads the lightning charge over high electrical (and thermal) conductivity copper wires, preventing excessive currents from running through and overheating the composite structure.

High *k* composite fibers may also be added to structure impacted by point heat sources, less severe than a lightning strike, to help spread the energy. While composite material systems normally have lower *k*-values compared to metals, carbon fibers are available with higher *k*-values than most metals.

References

1. John Manville, "Microlite® AA Premium NR Blankets," accessed April 30, 2017, https://www.jm.com/content/dam/jm/global/en/oem/OEM-data%20pages/Aerospace/AI-101_Microlite%20AA%20Premium%20NR.pdf.

2. Polymer Technologies Inc., "PolydampHydrophic Melamine Foam Ultra-Lite (PMF-UL)," accessed April 2017, https://www.polytechinc.com/products/polymer-acoustic-foam.

3. Solimide® Foams. Boyd Corporation, "HT-340 Product Data sheet," accessed April 2017, https://www.buckleyind.com/content/upload/files/Data%20Sheet-11-1.pdf.

4. Johns Manville, "Q-Fiber Felt Product Data Sheet," accessed April 2017, https://www.jm.com/content/dam/jm/global/en/oem/OEM-data%20pages/Aerospace/HPI-25_Q-Fiber%20Felt.pdf.

5. Morgan Advanced Materials, "Flexible Min-K® Datasheet," accessed April 2017, http://www.morganthermalceramics.com/media/4762/2-23-17-flexmink_mgam1.pdf.

6. Aspen Aerogels, "Pyrogel XTE Product Data Sheet," accessed April 2017, https://www.aerogel.com/_resources/common/userfiles/file/Data%20Sheets/Pyrogel-XTE-Datasheet.pdf.

7. IPS Innovative Products & Systems, "Colorful Coatings for Exterior Applications with Low Solar Absorption," accessed April 30, 2017, http://www.ips-innovations.com/solar_reflective_coatings.htm.

8. AIN Staff, "Qinetiq Sunscreens Aircraft," *Aviation International News*, December 13, 2006, accessed June 20, 2017, http://www.ainonline.com/aviation-news/aerospace/2006-12-13/qinetiq-sunscreens-aircraft?amp.

9. ArchToolbox, "Intumescent Paint, Fireproofing, and Firestopping," accessed April 30, 2017, https://www.archtoolbox.com/materials-systems/thermal-moisture-protection/intumescent-paint-fireproofing-and-firestopping.html.

10. Simonite, T., "I Moore's Law Is Dead. Now What?" *Intelligent Machines* (May 13, 2016), https://www.technologyreview.com/s/601441/moores-law-is-dead-now-what/.

11. Koomey, J., Berard, S., Sanchez, M., and Wong, H., "Implications of Historical Trends in the Electrical Efficiency of Computing," *IEEE Annals of the History of Computing* 33, no.3 (March 29, 2010): 46–54, doi:10.1109/MAHC.2010.28, ISSN 1058-6180.

12. Morgan, J., "Thermal Simulation and Testing of Expanded Metal Foils Used for Lightning Protection of Composite Aircraft Structures," *SAE Int. J. Aerosp.* 6, no. 2 (2013): 371-377, doi:10.4271/2013-01-2132.

Analytical Modeling

"All models are wrong, but some are useful."

—George Box [British statistician]

9.1 Introduction

Analytical modeling involves simplifying a system so that it may be represented by a series of equations that can be solved to predict performance. Modeling can predict structure and system responses to vehicle operational and outside ambient environments. The frequently reduced cost and time advantage of analytical modeling, compared to testing, and the ability to simulate dangerous conditions, such as fires, make it a critical design aid for *aircraft thermal management* (ATM).

The latest modeling tools, with their *graphical user interfaces* (GUIs), remove much of the drudgery of decades past when analysts calculated each individual node or element by hand, carefully chosen to minimize model runtimes. Expensive "mainframe" computers, with processors slower than the chips in today's smartphone, generated huge bills for the inefficient modeler using excessive nodes and elements creating slow converging models.

While computing time charges are largely a relic of the pre-*personal computer* (PC) era, understanding the basic concepts behind thermal and fluid flow systems modeling is still vital to ensuring credible modeling assumptions and error checking.

This chapter introduces the core modeling concepts and equations behind the thermal and fluid flow analytical simulations used in ATM.

9.2 Mathematical Modeling of Heat Transfer

Thermal analysts can simulate most heat transfer paths occurring throughout an aircraft with a few simple equations in *lumped capacitance networks*. Networks consist of *thermal*

©2020 SAE International

resistances to heat transfer; *thermal capacitances*, which store and release heat; and *forcing functions* (*boundaries*), which drive temperature change.

9.2.1 Thermal Resistances

Thermal resistances impede temperature changes (heat transfer) caused by *conductive*, *convective*, and *radiative* heat transfer.

9.2.1.1 CONDUCTION

Conductive heat transfer (conduction) occurs in solid materials from the transfer of kinetic energy as higher temperature (faster moving) particles strike slower moving (lower temperature) particles. The material temperature measures the kinetic energy change. The magnitude of conduction is determined by the thickness of the object in the direction of heat flow and the *thermal conductivity* (*k*) (ability to conduct heat) of the material.

$$q = \frac{kA}{L}(T_1 - T_2) \tag{9.1}$$

where
  q = heat transfer rate, Btu/hr (J/s)
  k = thermal conductivity, Btu/hr-ft-°F (W/m-K)
  A = cross-sectional area for heat transfer, ft^2 (m^2)
  $(T_1–T_2)$ = temperature difference within a solid, °F (°C)
  L = distance between temperature measurement locations, ft (m)

Thermal conductivity in a solid depends solely on the material and temperature (and the material composition for composites), and is measured using a simple test apparatus and procedure (Section 11.3.1). Composite materials are more complex to characterize then isotropic solids due to the thermal conductivity changes with fiber orientation. Isotropic materials have constant properties in all directions.

The combination of parameters multiplied by the temperature difference within the solid is called a *conduction thermal conductance* (G_k) or *conduction thermal conductor*.

$$G_k = \frac{kA}{L} \tag{9.2}$$

where
  G_k = conduction thermal conductance, Btu/h-°F (W/°C)

A *thermal resistance* (*R*) is the inverse of the G_k.

$$R_k = \frac{1}{G_k} = \frac{L}{kA} \tag{9.3}$$

where
  R_k = conduction thermal resistance, °F-h/Btu (°C/W)

Equations (9.2) and (9.3) assume a constant heat transfer area for the conductor path length. Solids where the heat transfer area varies along the flow length use an A/L term calculated by integrating the equation defining the area change versus distance (dA/dx).

A commonly used example is a cylinder, which can represent duct insulation covering a circular duct.

Cylinder thermal resistance

$$R_k = \frac{\ln(r_2/r_1)}{2\pi L k} \tag{9.4}$$

where
r_1 = cylinder inner radius, in (m)
r_2 = cylinder outer radius, in (m)
L = cylinder length, ft (m)

9.2.1.2 CONVECTION

Convective heat transfer (convection) is the process of heat flowing from hot to cold (through molecular interactions) when a fluid contacts a surface at a different temperature.

$$q = h\left(T_{surf} - T_\infty\right) \tag{9.5}$$

where
h = convective heat transfer coefficient, Btu/h-ft²-°F (W/m²-K)
T_{surf} = surface temperature, °F (°C)
T_∞ = bulk fluid temperature, °F (°C)

Convection operates like conduction with the warmer fluid or contacting surface transferring kinetic energy to the colder fluid or surface. Determining the convection conductor, specifically the convective *heat transfer coefficient* (*h-value*), however, is much more challenging then identifying the equivalent conduction property, *k*.

Convective *h-values* depend on complex hydrodynamic forces and multiple fluid and surface properties. The fluid *velocity* (*v*), *temperature* (*T*), *pressure* (*p*), *density* (*ρ*), *viscosity* (*μ*), *thermal diffusivity* (*α*), and the *surface orientation, flow distance* (*L*), *roughness* (*ε*), and overall contours, all influence convective heat transfer. Two of the most critical flow parameters, the *v* and impact angle, are also difficult to accurately estimate for aircraft internal surfaces. This causes a large variation in possible *h-values*, which empirically derived equations define using *dimensionless parameters* to characterize fluid surface interactions.

Dimensionless Parameters. Dimensionless parameters are ratios of fluid properties and surface characteristics where the units cancel out creating a number without units. They capture the different fluid and contacting surface characteristics affecting fluid flow and heat transfer. One of these parameters, the *h-value* containing *Nusselt number* (*Nu*), is calculated using other parameters (*Reynolds no.* (*Re*), *Prandtl no.* (*Pr*), *Grashof no.* (*Gr*), etc.), which are calculated from known inputs.

Nusselt Number. The *Nu* represents the ratio of convection to conductive heat transfer:

$$Nu_L = \frac{hL}{k_f} \tag{9.6}$$

where
L = characteristic length, ft (m)
k_f = thermal conductivity of a fluid, Btu/hr-ft-°F (W/m-°C)

L defines the fluid-to-surface contact dimension, which can represent a *length* (L) for flow over a flat plate or *diameter* (D) for internal flow or cross flow for a tube.

Heat transfer is entirely due to conduction when the Nu equals one.

Reynolds Number. The Re defines the ratio of inertial to viscous or friction forces within a fluid. Inertial forces refer to the resistance to change due to the mass of the fluid, while viscous forces (determined by the fluid viscosity) correspond to inter-molecular forces within the fluid resisting flow. Most fluids with high viscosity are "sticky."

$$Re_L = \frac{\rho v L}{\mu} = \frac{vL}{\gamma} \tag{9.7}$$

where

ρ = density, lbm/ft³ (kg/m³)
v = velocity, ft/s (m/s)
μ = dynamic viscosity, lbf-s/ft² (kg-s/m²)
γ = kinematic viscosity, ft²/s (m²/s)

The Re is the basis for identifying flow patterns in many different situations, from liquid in a pipe to air passing over an airplane wing, and for predicting the transition from *laminar* to *turbulent* flow.

Prandtl Number. The Pr represents the ratio of *kinematic viscosity* (γ), which is also called *momentum diffusivity*, to *thermal diffusivity* (α).

$$Pr = \frac{\gamma}{\alpha} \tag{9.8}$$

where
α = thermal diffusivity, ft²/s (m²/s)

Kinematic viscosity represents a fluid's resistance to flow with no external forces acting on it, outside of gravity. Thermal diffusivity is k divided by ρ and *specific heat at a constant pressure* (c_p), which represents how fast heat moves across a material.

The Pr equals about 0.7 for air, 0.7–1.0 for other gases, 1–10 for water, and 50–2000 for oils [1].

Grashof Number. The Gr represents the ratio of buoyancy to viscous forces. Differences in the density of a fluid, such as hot air rising or cold air sinking, generate buoyancy forces.

$$Gr_L = \frac{g\beta(T_{surf} - T_\infty)L^3}{\gamma^2}, \quad \beta = \frac{2}{(T_{surf} + T_\infty)} \tag{9.9}$$

where
β = volumetric thermal expansion coefficient, 1/°F (1/°C) for an ideal gas
g = acceleration due to gravity, ft/s² (m/s²)

Isaac Newton defined gravity as the attractive force two objects exert on each other, which is proportional to the mass of each and inversely proportional to the distance between them. The acceleration due to gravity on the surface of the earth is 32.174 ft/s² (9.807 m/s²) at sea level.

Contrary to popular belief, astronauts on the *International Space Station* (ISS) orbiting the earth at 248 mi (400 km) overhead do not face "zero gravity." The gravitational force is closer to 90% of the value faced on the surface of the earth. Instead, the station's 11,200 mph (18,000 km/hr) horizontal orbit creates a centrifugal force opposing the earth's gravitational pull. Thus, the ISS and astronauts are in a constant free fall, like astronauts training in an airplane that descends at the rate of *one gravity* (1-G) from a high altitude to provide a minute or so of weightlessness.

Rayleigh Number. Multiplying the *Gr* by the *Pr* generates the *Rayleigh number* (*Ra*), which characterizes free convection heat transfer conditions.

$$Ra = GrPr \qquad (9.10)$$

Flow Regimes. Choosing the most appropriate equation for a "real-life" scenario requires an understanding of the flow regime associated with specific dimensionless parameter values. Convection occurs over multiple categories including *forced* or *free*, and laminar or turbulent. Each category is based on observations indicating differences in the physical mechanisms affecting fluid flow and heat transfer. The dimensionless parameters used in the *h-value* equations also identify the flow type.

Forced versus Free Convection Heat Transfer. *Forced convection* occurs when a fluid is "forced" on to or across a surface by external means, such as a fan, pump, or the wind.

In the absence of a strong external force, temperature gradients in the fluid also cause movement as fluids rise and fall due to temperature driven ρ changes, generating *natural* or *free convection*. Fluid particles contacting warmer surfaces rise with an increasing temperature, while the same fluid contacting a colder surface sinks in response to a temperature decrease. Forced convection can also include free convection forces, generating mixed flow. Forced convection normally generates more heat transfer than free convection.

With the ECS responsible for moving air turned off, free convection occurs with a temperature difference between surfaces and the contacting air. This includes windows heated by the sun, or the warm surfaces of heat-generating equipment operated independently of the *electrical/electronic* (EE) cooling or *air-conditioning* (AC) system. With all ECS operating, free convection still occurs within passively cooled electronic boxes and racks, which include circuit boards represented by constant heat flux flat plates.

Outside the fuselage, free convection dominates the ullage and uncovered tank surface interface for fuel tanks lacking an inerting system. Supplying *nitrogen-enriched air* (NEA) may, however, generate sufficient air movement to create forced convection. Military and larger commercial aircraft have inerting systems to reduce fuel flammability and the risk of explosions.

Laminar versus Turbulent Flow. In laminar flow, fluid flows smoothly in layers parallel to a surface, at a constant speed, with no disruptions between the layers, while turbulent flow is irregular and unpredictable due to vortices and wakes. Both forced and free convection can be laminar or turbulent.

Small characteristic (flow) lengths and low ρ and high μ fluids flowing over smooth surfaces promote laminar flow. The driving factor for turbulence in forced convection is the fluid speed, while the temperature difference between the surface and contacting fluid has the same effect on free convection.

Flow starting out laminar at the leading edge of a surface (like a flat plate) can become turbulent as it moves along the surface in a forced or free convection regime.

The transition from laminar to turbulent flow occurs over a range of Re(s) and Gr(s) (a transition zone) starting with the critical Re_{cr} for forced and Gr_{cr} for free convection flow. The critical transition value varies with the surface shape and flow interactions. Therefore, the accepted 2300 Re_{cr} for forced convection flow in a smooth duct increases to 500,000 for forced convection flow over a smooth flat plate. These critical Re values can also vary widely in real life. Commercial airplane wings maintain laminar flow beyond the accepted theoretical Re_{cr} for a flat plate using innovative designs, like super-slippery paints or optimized contours. Reducing turbulence on aircraft external surfaces is important for reducing drag-generated fuel burn. Comparable variations in the Gr_{cr} can also occur in free convection flows.

While it takes impressive engineering to delay the onset of turbulence, the reverse, with turbulence occurring below the accepted Re_{cr} and Gr_{cr}, is more likely for flow scenarios affecting aircraft heat transfer. The variable velocity of wind is a typical forced convection turbulence promoter overlooked with a "cookbook" approach to *h-value* calculations.

Full turbulence in forced convection is reached around twice the Re_{cr} with the Nu gradually taking on the turbulent characteristics as the Re increases over the *transition zone*. Predicting the onset of turbulence is important because it can change convective heat transfer rates. For example, turbulent airflow from a gentle summer breeze cooling a sun-soaked black fuselage could significantly reduce measured skin temperatures during hot-day testing in Kuwait City. An analyst understanding this effect would use a different *h-value* correlation to provide a more accurate assessment of their model's performance.

Fully Developed Flow. The velocity of fluid entering a tube or contacting the leading edge of a surface slows near the wall due to shear forces in the fluid. The boundary layer thickens with increasing flow distance until it stabilizes, defining the start of *fully developed flow*. This can occur with either laminar or turbulent flow.

Boundary layer changes affect flow resistance and heat transfer, and should be evaluated using Nu correlations for the *thermal entrance length* occurring prior to the start of fully developed flow. The term fully developed flow often appears in references to closed-channel (duct) internal flow.

Constant Temperature versus Constant Heat Flux. Flow over a *constant temperature wall* may generate sufficient differences in convective *h-values*, compared to the same flow over a surface receiving a *constant wall heat flux*, to justify different correlations.

An airplane's external skin during flight provides a constant wall temperature since high external *h-values* ensure that internal aircraft heat sources (excluding failures like a fire) have a minimal influence on that boundary. A wire carrying electricity represents a constant heat flux surface, since the temperature adjusts (as needed) to dissipate a relatively uniform heat flux per unit length.

Choosing a Correlation. After identifying the flow regime, an analyst must choose a surface shape, used in empirical testing, with similar flow characteristics compared to the aircraft location. Testing completed for simple shapes, like flat plates and cylinders, can represent the complex shapes on an aircraft, based on engineering judgment.

Noncircular aircraft surfaces are divided among a series of flat plates, making this the most commonly used *h-value* correlation. Flat plates can even represent the skin's external surface of a moving fuselage due to the relatively small curvature and constant diameter aft of the flight deck. For an aircraft moving on the ground at a normal taxi speed or flying, fully developed forced convection flow occurs on the external constant diameter fuselage sections.

Forced Convection Correlations. With *environmental control systems* (ECS) operating, air movement throughout a fuselage generates forced convection heat transfer on planar surfaces represented by flat plate correlations. While the flow is rarely perfectly parallel to the surface, flat plate correlations are still appropriate for predicting *h-values*.

The *Re* and *Pr* define forced convection *Nu* correlations that often follow Equation (9.11).

$$Nu = CRe^m Pr^n \tag{9.11}$$

where

C, m, and n are flow-specific constant values in equations valid for a range of *Re* and *Pr*. Material property evaluations occur at the average temperature between the surface and fluid called the *film temperature* (T_{film}).

Flat Plate Flow. A local *Nu* (Nu_x) correlation generates *h-values* for surfaces located a distance x from the leading edge in contact with an air stream. An average *Nu* (Nu_L) calculates an average *h-value* for the length of the surface running in the laminar or turbulent flow region. Both *x* and *L* are the characteristic lengths used in calculating the dimensionless parameter (*Re*) used in the heat transfer correlation.

Laminar or Turbulent - Local or Average. Table 9.1 provides multiple versions of a flat plate forced convection *Nu* to accommodate a range of flow conditions [2].

TABLE 9.1 Flat plate forced convection heat transfer Nu for laminar and turbulent flow [2]

Constant wall temperature (isothermal)				
Flow regime		**Limits**	**Correlation**	**Eq.#**
Laminar	Local	$Re_x < Re_{cr}$, $0.6 < Pr < 50$	$Nu_x = 0.332 Re_x^{1/2} Pr^{1/3}$	9.12
	Average	$Re_L < Re_{cr}$, $Pr \geq 0.6$	$Nu_L = 2Nu_{x=L} = 0.664 Re_L^{1/2} Pr^{1/3}$	9.13
Turbulent	Local	$Re_{cr} < Re_x < 10^7$, $Pr \geq 0.6$	$Nu_x = 0.0296 Re_x^{4/5} Pr^{1/3}$	9.14
	Local	$10^7 < Re_x < 10^9$,	$Nu_x = 1.596 Re_x (\ln Re_x)^{-2.584} Pr^{1/3}$	9.15
	Average	$Re_{cr} < Re_L < 10^7$	$Nu_L = 0.037 Re_L^{4/5} Pr^{1/3}$	9.16
	Average	$10^7 < Re_L < 10^9$	$Nu_L = 1.967 Re_L (\ln Re_L)^{-2.584} Pr^{1/3}$	9.17
Constant wall heat flux				
Laminar	Local	$Re_x < Re_{cr}$, $0.6 < Pr < 50$	$Nu_x = 0.453 Re_x^{1/2} Pr^{1/3}$	9.18
Turbulent	Local	$Re_x > Re_{cr}$, $0.6 < Pr \leq 60.0$	$Nu_x = 0.0308 Re_x^{4/5} Pr^{1/3}$	9.19

$Re_{cr} = 500,000$

Mixed Flow (Laminar Plus Turbulent) - Average. Equations (9.20) and (9.21) combine the *average* laminar and turbulent flow *Nu* calculations (for a constant wall temperature) running from the leading edge to a distance *L* (past the transition zone) to capture *mixed flow* [3].

$$Nu_L = \left(0.037 Re_{cr}^{0.8} - 871\right) Pr^{1/3} \tag{9.20}$$

For $Re_L \leq 10^7$

$$Nu_L = \left[1.967 Re_L \left(\ln Re_L\right)^{-2.584} - 871\right] Pr^{1/3} \tag{9.21}$$

For $Re_L > 10^7$

Flow across a Cylinder.

The many cylindrical surfaces inside the aircraft, which include individual and bundles of wires, tubes carrying hydraulic fluids or liquid coolants, or air ducts, require a different correlation to predict convective *h-values*.

Laminar plus turbulent – heated cylinder

$$Nu_d = cRe_d^n Pr^{1/3}$$

(9.22)

where

Re#	Constant c	Constant n
0.4 – 4.0	0.989	0.33
4 – 40	0.911	0.365
40 – 4000	0.683	0.466
4,000 – 40,000	0.193	0.616
40,000 – 400,000	0.0266	0.805

Equation (9.22) calculates an (average) Nu_D for flow across a cylinder [4], which covers the smallest diameter wire and largest constant diameter commercial airplane fuselage cross section. While a fuselage exhibits large differences in heating between the sun and shaded surfaces, and not a constant heat flux, using Equation (9.14) is preferable to the alternative (guessing).

Flow along a Cylinder.

Nu correlations for air flowing parallel to a cylinder are mostly absent from heat transfer textbooks. Perhaps due to this flow scenario not representing a typical engineering design, unlike external cross flow which occurs in many industrial processes. Tube heat exchangers are a common example of cross flow with a fluid impinging on multiple closely spaced tubes designed to transfer heat.

Bond and Seban [5] address flow along a cylinder as an alternative to using a flat plate correlation.

Flow Inside a Tube.

Internal forced convection heat transfer occurs between fluid flowing inside a tube or duct and the contacting surface. On aircraft this covers fuel, hydraulic fluid, and liquid coolant flow through tubes, and air flowing through ducts. Both tubes and ducts have smooth surfaces to minimize the pressure drop.

Laminar - entrance plus fully developed - constant wall temperature - circular tube

$$Nu_d = 3.66 + \frac{0.0668\left(\dfrac{d}{L}\right)Re_d Pr}{1 + 0.04\left[\left(\dfrac{d}{L}\right)Re_d Pr\right]^{2/3}} \qquad (9.23)$$

where
   Nu_d = average Nusselt number over the duct length L
   d = duct diameter, in (m)
   L = duct length, in (m)

The transition to turbulence normally occurs around Re_d = 2300.

Convective heat transfer is highest at the tube entrance and decreases as the fluid moves along the tube, until becoming fully developed with a constant Nu_d. The thermal entrance length can be a significant part of the total *h-value* for laminar flow, making it a good practice to account for it using Equation (9.23) [6] rather than assuming the constant value associated with fully developed flow.

For a sufficiently long duct, the Nu_d approaches 3.66, the fully developed laminar flow value. A constant wall temperature is typically the most appropriate configuration for aircraft ducting installations. For a uniform heat flux, the fully developed Nu_d increases from 3.66 to 4.36. A constant wall temperature and uniform heat flux refer to the duct length and not time.

The thermal entrance length and Nu_d change in turbulent flow are usually small enough to ignore, so using the fully developed turbulent Nu_d, Equation (9.24) [7] correlation for the entire duct run is a reasonable approach.

Turbulent – fully developed – constant wall temperature – smooth circular tube

$$Nu_d = 0.023 Re_d^{0.8} Pr^n \qquad (9.24)$$

For 0.6 < Pr < 100 and 2500 < Re_d < 1.25 × 10⁵
where
   n = 0.4 for heating/n = 0.3 for cooling
   $_d$ means the tube diameter is the characteristic length.

A constant heat flux generates a similar Nu_d to a constant wall temperature, making a second equation unnecessary.

Free Convection Correlations. Free convection correlations are very similar to the forced convection, with the Gr replacing the Re of Equation (9.11).

$$Nu_{ave} = CGr^m Pr^n \qquad (9.25)$$

The Gr captures the effect of gravity-driven buoyant forces. Since the exponents m and n are usually identical, the Ra number replaces $GrPr$ in the free convection Nu correlations.

A fluid contacting a surface at a different temperature will either rise or fall unless prevented by the surface orientation. The fluid-to-surface temperature difference and surface orientation drive the fluid speed adjacent to the surface, and resulting convective heat transfer coefficient. Following are common free convection *h-value* correlations for a horizontal [8] and vertical [9] flat plate.

Horizontal Flat Plate. Equations assume isothermal wall (McAdams).
Upward facing hot or downward facing cold

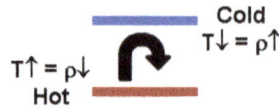

$$\overline{Nu_L} = 0.54\,Ra_L^{1/4}$$

(9.26)

For $2 \times 10^4 < Ra_L < 8 \times 10^6$

$$\overline{Nu_L} = 0.15\,Ra_L^{1/4}$$

(9.27)

For $8 \times 10^6 < Ra_L < 10^{11}$

where

$L = A/P$, ft (m)

A = area, ft^2 (m^2)

P = perimeter, ft (m)

Downward facing hot or upward facing cold

$$Nu_L = 0.27\,Ra_L^{1/4}$$

(9.28)

For $10^5 < Ra_L < 10^{11}$

Vertical Flat Plate. Laminar through turbulent flow, isothermal wall [9]

$$\overline{Nu_L} = \left\{ 0.825 + \frac{0.387\,Ra_L^{1/6}}{\left[1 + \left(0.492/Pr\right)^{9/16}\right]^{8/27}} \right\}^2$$

(9.29)

For $10^{-1} < Ra_L < 10^{12}$

Inclined Flat Plate. Equation (9.29) may be used for an inclined surface by replacing the Gr of equation (9.9) with equation (9.30) when calculating the Ra input parameter.

$$Gr_L = \frac{g\cos\phi\,\beta\left(T_s - T_\infty\right)L^3}{\gamma^2}$$

(9.30)

where

ϕ = surface angle from vertical, degs, $-60° < \phi < 60°$

g = acceleration due to gravity, 32.174 ft/s^2 (9.81 m/s^2) at sea level on earth

Mixed Convective Heat Transfer. A *combined* or *mixed Nu* captures the net result of the interaction between forced and free convection forces [10].

$$Nu_{combined} = \left(Nu_{forced}{}^{n} \pm Nu_{free}{}^{n} \right)^{(1/n)} \quad (9.31)$$

where

n = a constant determined by the geometry. The most used value is 3.
Nu_{forced} = the forced convection Nu
Nu_{free} = the natural or free convection Nu

Nu_{free} is added to Nu_{forced} when bouncy forces act in the same direction and subtracted when they act in the opposite direction as the forced convection (Figure 9.1).

The Nu_{free} in Figure 9.1 is additive for the hot side since air heated by the hot wall rises and subtracted for the cold wall since cold air sinks.

A *convection thermal conductance* (G_h) comparable to the G_k is generated using the calculated *h-value*.

$$G_h = hA \quad (9.32)$$

where

G_h = convection thermal conductance, Btu/h-°F (W/°C)

A *convective thermal resistance* (R_h) is the inverse of the G_h.

$$R_h = \frac{1}{G_h} = \frac{1}{hA} \quad (9.33)$$

where

R_h = convection thermal resistance, °F-h/Btu (°C/W)

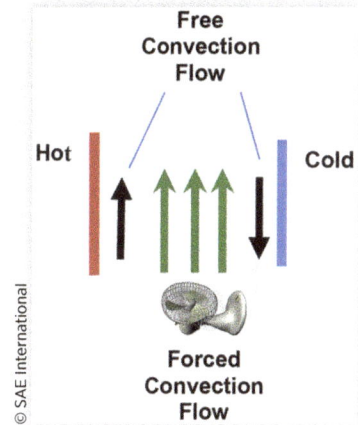

© SAE International

FIGURE 9.1 Mixed convective heat transfer.

9.2.1.3 RADIATION

All bodies above absolute zero, −460°F (−273°C), emit energy by electromagnetic radiation consisting of charged particles (protons and electrons) which can move through the vacuum of space and the gaseous atmosphere of the earth. *Radiative heat transfer* (radiation) occurs between surfaces as these charged particles released from both surfaces strike the other surface.

$$q = A\sigma\Im\varepsilon \left(T_1^4 - T_2^4 \right) \quad (9.34)$$

where

σ = Stefan-Boltzmann constant, a factor which converts temperature to heat 1.714×10^{-9} Btu/hr · ft²-°R⁴ (5.6704×10^{-8} W/m² · K⁴)
$\Im$ = shape, or view factor between two surfaces, 0–1.0
ε = emissivity, a measure of the ability of an object to emit radiative energy, 0–1.0

While conduction and convection are linearly dependent on temperature, radiation is proportional to the absolute temperature to the fourth power. Linearizing radiation makes it compatible with conduction and convection, to support the creation of thermal networks to solve by hand or with a computer program.

$$\left(T_1^4 - T_2^4 \right) = \left(T_1 + T_2 \right)\left(T_1^2 + T_2^2 \right)\left(T_1 - T_2 \right) \quad (9.35)$$

CHAPTER 9

Substituting Equation (9.35) into Equation (9.34)

$$q = \left\{ A \sigma \Im \varepsilon \left(T_1 + T_2 \right) \left(T_1^2 + T_2^2 \right) \right\} \left(T_1 - T_2 \right) \tag{9.36}$$

Removing the temperature difference term from (9.36) creates a linear *radiation thermal conductance* (G_r).

$$G_r = \left\{ A \sigma \Im \varepsilon \left(T_1 + T_2 \right) \left(T_1^2 + T_2^2 \right) \right\} \tag{9.37}$$

where
G_r = radiation thermal conductance, Btu/h-°F (W/°C)

The *radiation thermal resistance* (R_r) is the inverse of G_r.

$$R_r = \frac{1}{A \sigma \Im \varepsilon \left(T_1 + T_2 \right) \left(T_1^2 + T_2^2 \right)} \tag{9.38}$$

where
R_r = radiation thermal resistance, °F-h/Btu (°C/W)

9.2.2 Thermal Capacitance

Thermal capacitance (C) is a measure of how much energy an object can store.

$$C = mc_p \tag{9.39}$$

where
C = thermal capacitance, Btu (J)
m = mass, lbm (kg)
c_p = specific heat, Btu/lbm-°F (J/kg- C)

Specific heat is the heat needed to raise a unit mass of material by a given amount.

9.2.3 Energy Sources

Thermal models usually include energy sources which cause heat flow and temperature changes. Examples of energy sources are a heat load on a piece of equipment like an avionics box, a boundary temperature like the outside ambient air, or fluid flow that can cool or warm a compartment, equipment, or structure.

9.2.4 Mass Transfer (Fluid Flow)

Fluid flowing into a volume of a different temperature transfers heat per Equation (9.40).

$$q_{in} = \dot{m} c_p \left(T_{in} - T_\infty \right) \tag{9.40}$$

where
q_{in} = the amount of heat entering a system due to fluid flow, Btu/h (W)
$\dot{m}$ = mass flow, lbm/h (kg/s)
T_{in} = temperature of fluid entering a control volume, °F (°C)
T_∞ = temperature of fluid in control volume, °F (°C)

9.2.5 Analytical Modeling Using the Electrical Analogy

Conductive heat transfer through a solid, Equation (9.1), is like the flow of electricity in a wire, leading to an *electrical analogy*, based on the basic circuit, Equation (9.41).

$$I = \frac{\Delta V}{\sum R_{elec}}$$ (9.41)

where

I = electrical current, amps
$\sum R_{elec}$ = summation of wire electrical resistances, W/amp^2 (ohms)
ΔV = voltage difference, W/amps (volts)

Equation (9.42) shows the thermal equivalent for the electrical circuit equation.

$$q = \frac{\Delta T}{\sum R}$$ (9.42)

where

ΔT = temperature difference, °F (°C)
$\sum R$ = summation of linear thermal resistances, °F-h/Btu (°C/W)

In this analogy, heat flux (q), represents current (I), temperature difference (ΔT) corresponds to a voltage difference (ΔV), and linear thermal resistances ($\sum R$) represent electrical resistances ($\sum R_{elecc}$). A temperature difference drives heat transfer while a voltage difference drives the flow of electricity (current). A thermal circuit is, therefore, solved by looking at the sum of the energy entering and leaving each lumped thermal node. The key requirement for the *thermal electrical analogy* equation is being able to define the heat transfer using a linear temperature difference.

9.2.5.1 SERIES RESISTANCE

Adding multiple resistances in series, like layers of clothing for a Minnesotan heading out for an evening stroll in January, creates a single equivalent thermal resistance.

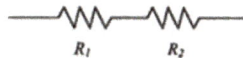

$$R_1 + R_1 = R_{12}$$ (9.43)

9.2.5.2 PARALLEL RESISTANCE

Parallel resistances, like heat transfer through a wall with a window in the center, can also be reduced to a single equivalent resistance.

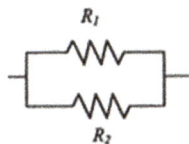

$$\frac{1}{R_1} + \frac{1}{R_2} = \frac{1}{R_{12}}$$ (9.44)

Which is equivalent to

$$R_{12} = \frac{R_1 R_2}{R_1 + R_2} \tag{9.45}$$

For three parallel resistances

$$R_{123} = \frac{R_1 R_2 R_3}{R_1 R_2 + R_2 R_3 + R_3 R_1} \tag{9.46}$$

9.2.5.3 EXAMPLE: HEAT TRANSFER FROM AN INSULATED HOT-AIR DUCT

Thermal resistances added in series and parallel support the calculation of performance parameters, such as the heat rejection from an insulated thin metal duct carrying hot air (Figure 9.2).

FIGURE 9.2 Insulated hot-air duct thermal network.

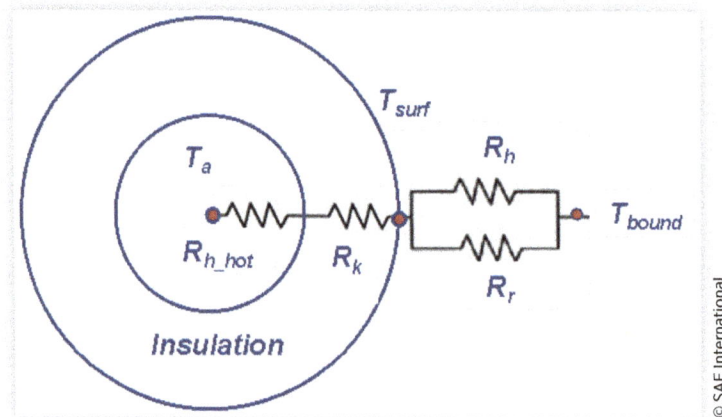

© SAE International

Since metals have a very high k relative to insulation, we can ignore their thermal resistance. For this example, the air and surface temperatures of adjacent structure (T_{bound}) are identical leading to the following mathematical representation.

$$q = \frac{(T_a - T_{bound})}{R_{h\_hot} + R_k + \dfrac{R_h R_r}{R_h + R_r}} \tag{9.47}$$

where

T_a = hot air temperature, °F (°C)
T_{bound} = surrounding air and surface temperatures, °F (°C)
R_* = thermal resistance corresponding to a heat transfer mode, °F-h/Btu (°C/W)
$_k$ = insulation conduction
$_h$ = insulation surface convection
$_{h\_hot}$ = internal duct surface convection
$_r$ = insulation surface radiation

Boundary conditions required to solve this problem include

- *Hot-air* flow rate, temperature, and pressure
- *Duct* diameter, length, and surface roughness
- *Insulation* thickness, thermal conductivity, and outer emissivity
- *Surrounding air/surface* temperatures and flow characteristics

All thermal resistances can be calculated with the addition of the insulation surface temperature. The steady-state insulation temperature is determined iteratively by varying the *insulation surface temperature* (T_{surf}) in Equations (9.48) and (9.49) until heat entering the insulation from the hot air equals heat leaving it from the outer surface.

$$\frac{\left(T_a - T_{bound}\right)}{R_{h\_hot} + R_k + \dfrac{R_h R_r}{R_h + R_r}} - \frac{\left(T_a - T_{surf}\right)}{R_{h\_hot} + R_k} = 0 \tag{9.48}$$

$$R_r = \frac{1}{\left(A\sigma\Im\varepsilon\right)_{surf}\left(T_{surf} + T_{bound}\right)\left(T_{surf}^{2} + T_{bound}^{2}\right)} \tag{9.49}$$

9.2.5.4 ITERATIVE METHOD

The process of choosing a new more accurate T_{surf} based on the error generated by a previous guess is called an *iterative method*. The error in this instance is the heat accumulated by the insulated surface node per Equation (9.48).

Multiple methods are available for choosing the new temperature based on the error, including the Gauss-Seidel, Jacobi, and Bisection Methods discussed in numerical analysis textbooks. A numerically inefficient, but quick, way to implement an iterative method involves varying the temperature based on the magnitude of the error generated using the simple equation (9.50).

$$\left(T_{surf}\right)_{New} = \left(T_{surf}\right)_{Old} + \frac{Q_{error}}{Factor} \tag{9.50}$$

where

$_{New}$ = the new temperature guess
$_{Old}$ = the prior calculated temperature
Q_{error} = net heat gain at surface, Btu/h (W)
Factor = relaxation factor used to choose new insulation surface temperature (number determined by trial and error)

Starting with a larger *Factor*, which causes a slow convergence, is usually more efficient than starting too small and risking a solution divergence that can quickly generate a divide-by-zero error. Divide-by-zero errors in Excel are especially time consuming to repair.

9.3 Mathematical Modeling of Airflow Systems

Systems that move fluids from one location to another can also be modeled using a *one-dimensional* (1-D) network approach, similar to a thermal network.

While temperature differences drive heat flow through thermal resistances, pressure differences drive fluid through networks of flow resistances. The basis of 1-D fluid flow analysis starts with *Bernoulli's equation.*

9.3.1 Bernoulli's Equation

Swiss mathematician and physicist Daniel Bernoulli applied mathematics to fluid mechanics developing the concept of *Conservation of Energy* which states that energy can neither be created nor destroyed in a closed system, and is therefore constant.

$$p + \frac{\rho v^2}{2g_c} + \frac{\rho gz}{g_c} = \text{constant} \tag{9.51}$$

where

p = pressure at the chosen point, lbf/ft^2 (Pa or N/m^2)
ρ = density of the fluid, lbm/ft^3 (kg/m^3)
v = velocity of the fluid, ft/s (m/s)
g = acceleration due to gravity, 32.174 ft/s^2 (9.81 m/s^2) at sea level on earth
g_c = gravitational constant, 32.174 lbm-ft/lbf-s^2 (set to 1 for SI units)
g_c is used for English units only to convert lbm to lbf (Section 2.2.3)
z = the elevation at the reference plane, ft (m)

The basic form of Bernoulli's equation is modified for specific applications by multiplying or dividing all terms by a physical parameter. This facilitates designing or analyzing systems for moving fluids or surfaces moving through fluids. These modifications accompany new terminology, like *head.*

9.3.1.1 HEAD

In hydraulics engineering, which involves liquids, dividing Equation (9.51) by ρg converts the different forms of energy to an equivalent height or head.

$$\frac{pg_c}{\rho g} + \frac{v^2}{2g} + z = \text{constant} \tag{9.52}$$

The quantities $pg_c/\rho g$ and $v^2/2g$ are the *pressure head* and *velocity head*, respectively, while z equals the *elevation head.* The pressure head represents the internal energy of the fluid from pressure exerted on its containment structure. Velocity head relates to the kinetic energy of movement, and elevation head relates to the energy associated with height differences (when dealing with a liquid).

9.3.1.2 GASES

Gases exclude the height component z because of its negligible impact on pressure loss with low ρ fluids. Representing the energy as pressures instead of height or head, therefore, makes more sense for gases.

$$\frac{\rho v^2}{2g_c} + p_s = \text{constant} \tag{9.53}$$

Equation (9.53) is used to calculate airflow in a ducting system by measuring a pressure drop, or aircraft speed using a pressure rise from a *pitot tube* attached to an airplane skin (Section 11.5.2.1). The combination $\rho v^2/2$, which is the *velocity or dynamic pressure* (p_v), is equivalent to the velocity head for hydraulic (liquid) flow. The remaining *static pressure* (p_s) represents the pressure on the duct with no air movement. The static plus velocity pressure equals the *total pressure* (p_t).

$$p_t = p_v + p_s \qquad (9.54)$$

9.3.2 Pressure Generation: Fans and Pumps

Both fans moving a gas (usually air) and pumps moving a liquid convert the kinetic energy of an impeller to a pressure to move a fluid. While the physics of their operation is identical, the method of defining pump and fan performance can vary slightly. Pumps are defined by the height, they can raise a fluid (called head), while fans are defined by the pressure they produce, in addition to an equivalent head. The choice of head or pressure makes sense after considering differences in liquid and gas properties and the application of pumps and fans.

9.3.2.1 FAN PRESSURE

A fan pressure rise defined as a force per unit area in analytical models is often presented as an equivalent head in fan performance curves and maps. The head height is based on a fluid, usually water or mercury, which are also used in manometers (Section 11.5.2.2) measuring pressure changes in lab testing.

Fan Performance Curves. Curves showing the pressure generated and power consumption (*brake horsepower*) versus volumetric flow define fan performance (Figure 9.3).

FIGURE 9.3 Fan performance curve.

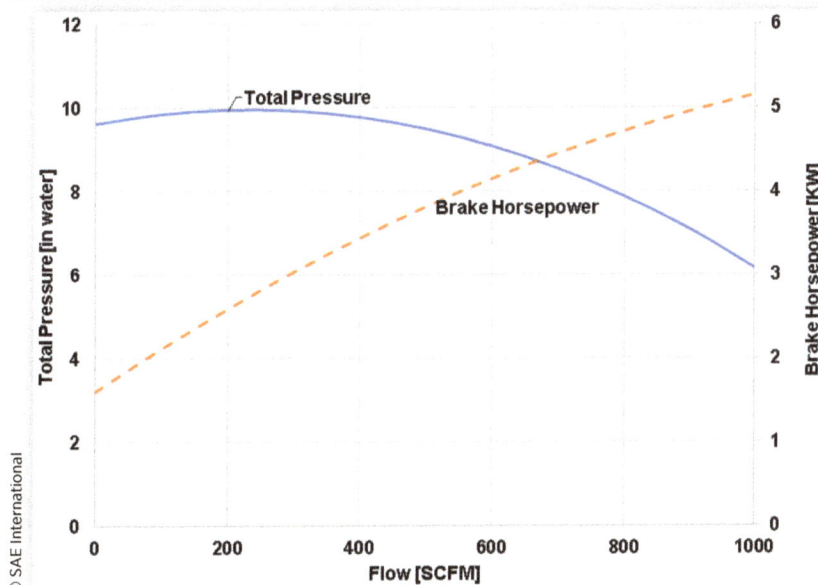

© SAE International

CHAPTER 9

Using head to describe a pressure for a fan can improve the chart readability by forcing the y-axis to a more convenient scale. Table 9.2 shows that switching from lbf/in^2 to in water increases the units almost 28 times. The performance curve from a fan generating 0.5 lbf/in^2 peak pressure is easier to read in inches of water running from around 0 to 14 in increments of 1 rather than fractions of 0.5.

TABLE 9.2 Pressure units

English units				SI units	
lbf/in^2	lbf/ft^2	in water	in Hg	Pa	kPa
1	144	27.71	2.036	6895	6.9

© SAE International

Fan performance curves may also include lines showing system pressure drop (impedance) versus volumetric flow rate (Figure 9.4).

FIGURE 9.4   Fan performance curve with system line.

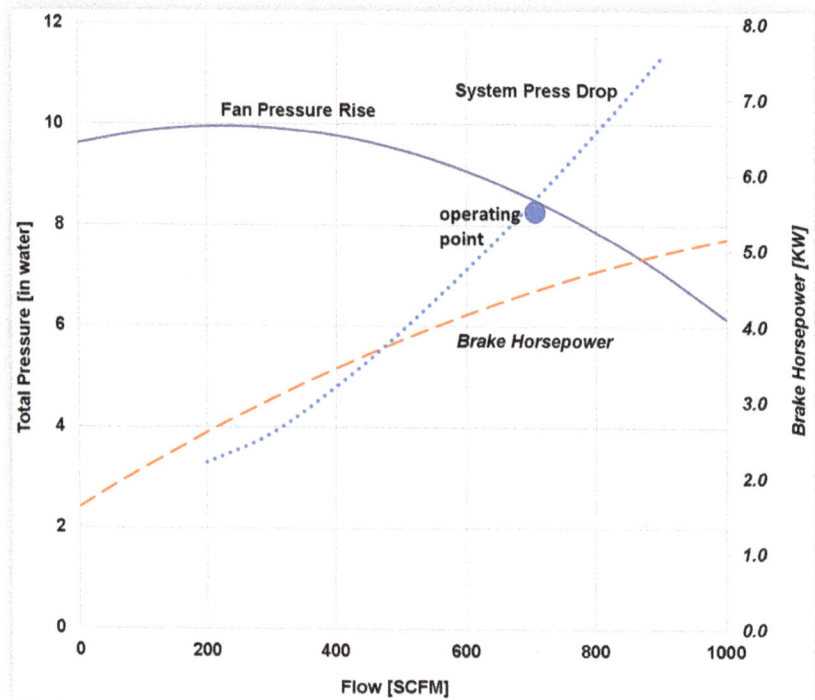

© SAE International

The intersection of the fan pressure generation line with the system pressure drop line provides a stable performance point for the system. Increases or decreases in the fan flow affect the system pressure drop by driving the fan flow rate back to the stable operating point.

9.3.2.2 PUMP HEAD

One of the earliest uses for pumps (which continues today) is raising water to a higher elevation to provide the pressure (from gravity) needed to overcome pressure loses in the pipes or canals serving a community or individual user. Rating pumps by their head versus flow rate eliminates the need for converting between units.

Pump Performance Curves. A pump performance curve provides similar information as a fan curve using different units while adding the efficiency, which may be missing from fan curves (Figure 9.5).

FIGURE 9.5 Pump performance curve.

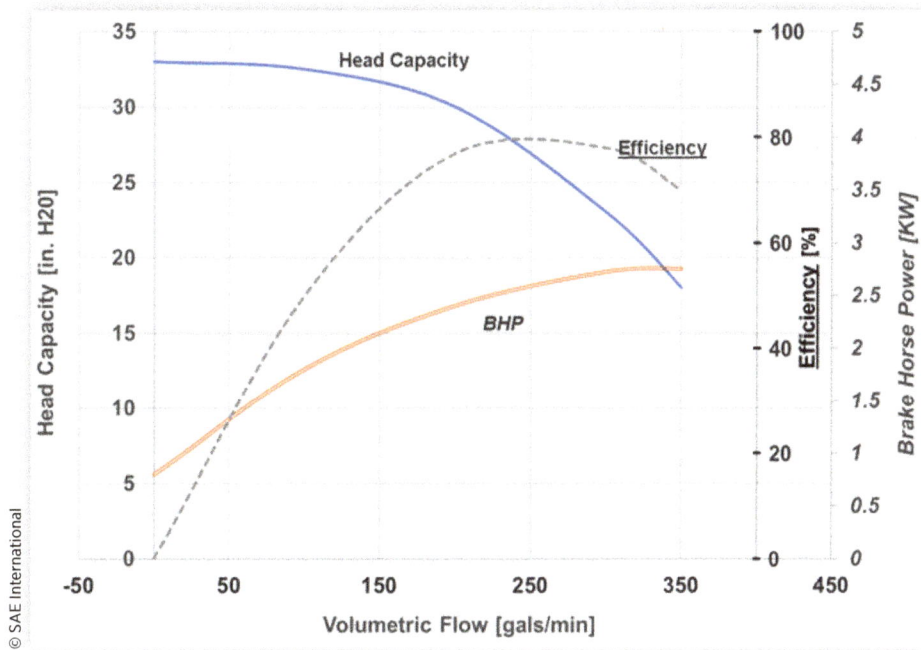

Pump and fan curves use the same terms for power consumption and efficiency, while height is always associated with a liquid.

9.3.3 System Pressure Drop

Fluid flowing through a closed conduit ducting or piping system must overcome *frictional*, *kinetic*, and for liquids, *hydrostatic* pressure losses. Frictional losses depend on the fluid properties and the pipe or duct material *roughness* (ε). Kinetic energy losses relate to flow disruptions resulting from changes in the direction and velocity of the fluid flow. This occurs when a fluid stream turns, diverges, converges, enters, exits, or passes through orifices, valves, filters, or fittings that impede fluid movement. Hydrostatic losses occur from raising the height of a liquid. Additional energy and pressure losses occur as fluid flows through heat exchangers, electronic boxes, or other equipment.

System component and frictional pressure losses are defined relative to the velocity head, and added in series to determine the minimum required fan or pump pressure rise.

$$\Delta p_t = \left\{ \Sigma \left(\frac{f_D L}{D_h} \right) \left(\frac{\rho v^2}{2} \right) + \Sigma K \left(\frac{\rho v^2}{2} \right) \right\} \quad \text{SI Units} \tag{9.55}$$

$$\Delta p_t = \left\{ \Sigma \left(\frac{f_D L}{D_h} \right) \left(\frac{\rho v^2}{2 g_c} \right) + \Sigma K \left(\frac{\rho v^2}{2 g_c} \right) \right\} \quad \text{English Units} \tag{9.56}$$

where

Δp_t = system total pressure drop, lbf/in^2 (Pa)

f_D = Darcy-Weisbach friction factor (dimensionless) (Section 9.3.3.1)

L = duct length, ft (m)

K = loss coefficient (dimensionless)

D_h = duct hydraulic diameter, ft (m)

For noncircular ducts $D_h = 4 * A/P$

where

A= cross-sectional area, ft^2 (m^2)

P = wetted perimeter, ft (m)

Frictional losses are calculated relative to the duct *length* (L) and *hydraulic diameter* (D_h), while the remaining system losses are calculated based on pressure *loss coefficients* (*k-values*).

9.3.3.1 FRICTIONAL LOSSES

Frictional losses caused by shear forces occurring at a duct internal surface can be evaluated by a dimensionless parameter called the *Darcy-Weisbach friction factor* (f_D). Some textbooks use the *Fanning friction factor* (f_f), while both equations yield the same results since $f_D = 4f_f$.

In 1944 Lewis Moody released his famous Moody chart, providing a graph of the f_D versus Re for fully developed flow in a pipe over a range of surface roughness values and Re (Figure 9.6).

The Moody chart includes laminar and turbulent flow regimes with a small transition region. The change in pressure loss going from laminar to turbulent flow can lead to instabilities in the system flow, which better designs avoid.

FIGURE 9.6   Moody Diagram [11].

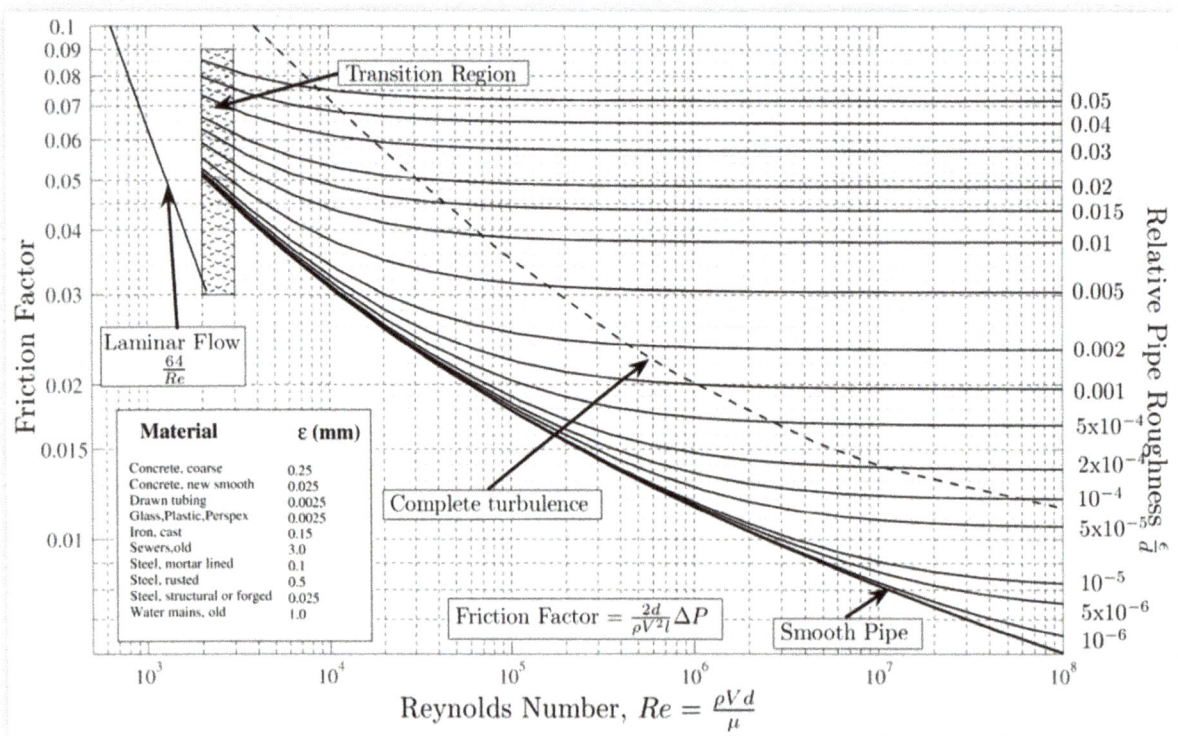

© Donebythe secondlaw/CC BY-SA 3.0

The *laminar flow* f_D (f_{D_lam}) is simply 64 divided by the *Re*.

$$f_{D\_lam} = \frac{64}{Re_D} \tag{9.57}$$

The *Colebrook-White* equation [12] is a well-known correlation for characterizing the turbulent flow friction factors (f_{D_turb}) listed on the Moody chart.

$$\frac{1}{\sqrt{f_{D\_turb}}} = -2\log_{10}\left(\frac{\varepsilon / D_h}{3.7} + \frac{2.51}{Re\sqrt{f_{D\_turb}}}\right) \tag{9.58}$$

where

ε = duct absolute surface roughness (m)

Donald Miller presents in *Internal Flow Systems* [9-13] a closed form equation that closely matches the accuracy of the Colebrook-White equation.

$$f_{D\_turb} = \frac{0.25}{\left[\log_{10}\left(\dfrac{\varepsilon}{3.7D_h} + \dfrac{5.74}{Re^{0.9}}\right)\right]^2} \tag{9.59}$$

Internal Flow Systems is an industry standard network fluid flow textbook that Mentor Graphics FloMaster (Section 10.3) is based on.

9.3.3.2 LOSS COEFFICIENTS

A dimensionless *loss coefficient* (*k-value*) defines fluid flow resistance as the fraction of the velocity pressure (kinetic energy of the flow).

$$k = \frac{\Delta p}{\rho v^2 / 2g_c} \tag{9.60}$$

Published *k*-value data comes in different forms. The simplest data consists of a single data point, like a *sharp-edged inlet pipe flush with a wall* with no limitations beyond a straight entrance length much longer than the flow diameter [14].

Sharp-edged Inlet pipe

K=0.5

Other *k-values* include tables of data for components, such as a *sharp-edged orifice* used to balance the branch flows in an air distribution or EE cooling system [15] (Table 9.3).

TABLE 9.3 Sharpe-edged orifice in a duct loss coefficient [15]

d/D	0.20	0.25	0.30	0.35	0.40	0.45	0.50	0.55	0.60	0.65	0.70	0.75	0.80	0.85	0.90
K	65	39	27	19	14	10	7.7	5.8	4.2	3.1	2.3	1.5	0.97	0.55	0.26

For calculations on a spreadsheet or MATLAB, an equation form is more convenient than the tabular data, such as the following duct entrance *k-value*.

Duct entrance flush with a wall at an arbitrary angle [16].

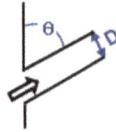

$$K = 0.5 + 0.3\cos\theta + 0.2\cos^2\theta$$

Conflicting Data. Analysts often must choose between significantly different *k-values* published from multiple sources, such as the rounded duct entrance at a flush wall interface data of Table 9.4 [17].

Table 9.4 shows a common problem of determining the most accurate value when multiple sources are available. The safest approach is to use the data driving the system size. This normally means using the larger *k-value* requiring a larger pressure rise to overcome.

TABLE 9.4 Rounded duct entrance at a flush wall interface duct loss coefficient [17]

d/D	0.00	0.01	0.02	0.03	0.04	0.05	0.06	0.08	0.12	0.16	>0.16	
K	0.50	0.43	0.36	0.31	0.26	0.22	0.20	0.15	0.09	0.06	0.03	(Idelchik)
K	0.44	0.35	0.28	0.22	0.17	0.13	0.10	0.07	0.03	0.00	0.00	(Harris)

Interaction between Closely Spaced Components. Duct pressure loss testing normally occurs with isolated components. Long straight ducts, attached to both ends of the component ensure a uniform flow velocity over the entire flow area at the measurement location. Flow fields from two components close to each other, however, can change the pressure drop characteristics of both components. The degree of interaction depends on the pipe or duct *span length* (L) between components and D_h. Blevins [18] identifies a *pressure loss interaction coefficient*, for two components, showing that interactions may significantly impact the pressure drop through the corresponding components for an L/D<30 (Table 9.5).

TABLE 9.5 Pressure loss interaction coefficients [18]

Length of intermediate spacer	$\frac{L}{D} > 30$	$30 > \frac{L}{D} > 10$	$10 > \frac{L}{D} > 3$	$\frac{L}{D} < 3$
Pressure loss interaction coefficient	1.0	0.7 to 1.0	0.5 to 1.0	0.5 to 3.0

Neglecting ducting component interactions can result in large errors in the pressure drop and accompanying flow predictions. This is a common situation since short component spacings are unavoidable for ducting systems located in a space-constrained aircraft. Recognizing this effect, researchers have measured pressure drops across multiple components close enough to interact, such as the two 90° bends in series (Table 9.6) [19].

Table 9.6 values are multiplied by the individual component loss coefficients, such as the 90° bend data (Table 9.7) [20], to predict a more accurate pressure drop.

TABLE 9.6 Interaction loss coefficient for 90° bends

	$K_{combination}/\Sigma K_{components}$				
	Spacing between components L/D				
R/D	0	4	10	20	30
1.85	0.86	0.72	0.82	0.95	0.96
3.3	0.84	0.82	0.86	0.96	1.00
7.5	0.93	0.96	0.97	1.00	1.00

© SAE International

TABLE 9.7 90° Smooth bend loss coefficient

R/D	1	2	4	6	8	10
K	0.35	0.19	0.16	0.21	0.28	0.32

© SAE International

Knowing to multiply the sum of the two 90° bend loss coefficients, which assume no interactions, by the values in Table 9.6, is one sign of a skilled analyst. Someone who pays attention to details while others depend on "fudge" (error) factors to match test data to one or two test conditions, claiming to "correlate" their model.

Some commercial network fluid modeling software can predict component interaction using a *computational fluid dynamics* (CFD) solver to simulate nonuniform flow velocities (Section 10.3).

***k-value* Data References.** Idelchik's *Handbook of Hydraulic Resistance* [14] provides an extensive collection of *k-value* test data corresponding to every imaginable hydraulic system component. The Soviet Union funded Idelchik's work for their atomic industry, and the US Atomic Energy Agency translated his book into English. Enthusiasm for this amazing resource may, however, be slightly tempered due to nomenclature differences. Fortunately, quality fluid dynamics handbooks by Blevins [15] and Miller [16] provide an abundance of *k-values* along with clearer explanations of fluid flow concepts. If you cannot find an appropriate *k-value* in one of the three books, it is probably not available to the general public.

9.3.4 Flow Calculation

Rearranging Bernoulli's equation to solve for the flow rate using the *k-values* and f_D data leads to Equation (9.61).

$$\dot{m} = \frac{A(2\rho g_c \Delta P)^{1/2}}{\left\{\left[\Sigma\left(\frac{f_D l}{D_h}\right) + \Sigma K\right]\right\}^{1/2}} \qquad (9.61)$$

where
    $\dot{m}$ = mass flow rate, lbm/s (kg/s)

Since the f_D is dependent on the Re, which is proportional to the flow rate, an iterative process is required to solve for the mass flow rate. This is similar to the method discussed to calculate the heat transfer from an insulated duct equation (Section 9.2.4), starting with a velocity guess.

References

1. Engineering ToolBox, "Prandtl Number," accessed May 3, 2017, http://www.engineeringtoolbox.com/prandtl-number-d_1068.html.

2. Suryanarayana, N.V., Section 3.2.2: Forced Convection—External Flows, *CRC Handbook of Thermal Engineering*, 3rd ed., Frank, K. ed. (New York: CRC Press, 2000), 3-26–3-46.

3. Ibid.

4. Manohar, K. and Ramroop, K., "Skin Friction and Heat Transfer Characteristics of a Laminar Boundary Layer on a Circular Cylinder in Axial Incompressible Flow," *International Journal of Engineering (IJE)* 4, no. 4 (2010): 268–278.

5. Bond, R. and Seban, R.A., "Skin Friction and Heat Transfer Characteristics of a Laminar Boundary Layer on a Circular Cylinder in Axial Incompressible Flow," *Journal of Aerospace Science* 18 (1951): 671–675.

6. Holman, J.P., *Heat Transfer Tenth Edition* (Boston, MA: McGraw-Hill, 2010), 312.

7. Ibid.

8. White, F.M., *Heat Transfer* (Reading, MA: Addison-Wesley, 1984), 345.

9. Holman, 2010, 334.

10. White, 1984, 355.

11. Wikipedia, "Moody Chart," 2017, accessed May 14, 2017, https://en.wikipedia.org/w/index.php?title=Moody_chart&oldid=747975983#/media/File:Moody_diagram.jpg.

12. Rollman, P. and Spindler, K., "Explicit Representation of the Implicit Colebrook–White Equation," *Case Studies in Thermal Engineering* 5 (2015): 41.

13. Miller, D.S., *Internal Flow Systems*, 2nd ed. (BHR Group, 1996), 191, ISBN-10: 0947711775.

14. Blevins, R.D., *Applied Fluid Dynamics Handbook* (Van Nostrand Reinhold Company, 1992), 72.

15. Ibid, 76.

16. Ibid, 72.

17. Ibid, 73.

18. Ibid, 70.

19. Ibid, 68.

20. Ibid.

21. Idelchik, I.E., *Handbook of Hydraulic Resistance*, 3rd ed. (Boca Raton, FL: CRC Press, 1994).

22. Blevins, 1992.

23. Miller, 1996.

10

Analytical Software

"Testers don't break software, software is already broken."

—Amir Ghahrai

10.1 Introduction

The wide range of physical phenomena affecting *aircraft thermal management* (ATM) is modeled using different types of analytical tools. The most common tools include *thermal* or *thermal/fluid systems*, *one-dimensional (1-D) network flow*, *computational fluid dynamics* (CFD), and *systems integration (multi-domain)* simulation programs. Many structural analysis programs also include thermal analysis solvers, which let users generate temperatures of nodes in their structural models. Much overlap occurs between these categories with some 1-D network flow simulation programs offering limited CFD modeling capabilities, and system integration tools including capabilities associated with the other analytical programs. A few large companies offer suites of tools covering all categories.

Existing tools are continually evolving, while new ones arrive on the market yearly, so the software mentioned does not imply preferred products.

10.2 Thermal/Fluid Systems

The primary use of thermal/fluid systems simulation programs is predicting structure, equipment, and fluid systems temperatures. They are frequently used to model an entire vehicle, while other analytical tools, such as CFD and network flow simulation programs, calculate thermal model input data. Thermal simulation programs often include fluid flow modeling capabilities of various levels. The most primitive can handle single-phase flow through a simple duct network, while the more sophisticated programs handle two-phase flow and include extensive component libraries.

10.2.1 **SINDA**

For the past 60 years, a version of the lumped capacitance thermal analysis code, the *Systems Integrated Numerical Differencing Analyzer* (SINDA), has been the aerospace standard thermal analysis program. SINDA is a legacy of the US space program with the US *National Aeronautics and Space Administration* (NASA) funding the development and making its use a requirement for all thermal modeling supporting government-funded spacecraft. SINDA also includes a fluid solver, FLUINT (short for FLUid INTegrator), leading to the name SINDA/FLUINT. Today, C&R Technologies [1] markets a commercial version of SINDA/FLUINT, while MSC Software offers their own SINDA-type code [2] and comparable fluid simulation program.

C&R Technologies is a small company founded by two of the primary developers of SINDA/FLUINT, while MSC is a large company offering a comprehensive set of analytical tools acquired over the years. Both company's products include extensive user enhancements, like *graphical user interfaces* (GUIs), radiation conductor generators, and advanced fluid modeling capabilities.

10.2.2 **Computer-Aided Design (CAD) Embedded**

Most *computer-aided design* (CAD) software includes at least basic thermal analysis capabilities, which allow the designer to predict structure temperatures using meshes created for other uses. This is a popular approach for equipment suppliers since it saves the time and cost to build a separate thermal model and eliminates the need for thermal analysis expertise (in the eyes of some).

This seemingly cost-efficient approach, however, can add technical risk, and even costs to a program. A model mesh optimized (using the minimum number of nodes) to produce accurate stress predictions may generate excessive errors when used for temperature predictions. On the other hand, a finely meshed model with far more nodes than needed for accurate temperature predictions can produce excessive runtimes. In addition, applying the theoretical boundary assumptions built into these programs, with no modifications to accommodate real-life understandings, can produce large errors in predictions. Choosing the appropriate convective *heat transfer coefficient* (*h-value*) for a complex shape is a challenging task which most stress analysts are inadequately prepared for. Determining thermal model boundary conditions often takes more engineering judgment compared to most stress analysis.

10.3 **1-D Network Flow**

One-dimensional network flow solvers model airplane systems that move a fluid from one location to another (Section 9.3). They are available as an add-on to analysis packages designed for a different purpose or as stand-alone codes.

The *FLUINT* fluid module in SINDA/FLUINT is a well-known example of an add-on flow solver for thermal analysts in the aerospace industry (Section 10.2.1).

MSC Software's *Easy5* [3] flow modeling function is another example of an add-on capability since Boeing originally wrote the code for controls analysis, before adding extensive *environmental control systems* (ECS), hydraulics, and fuel modeling capabilities and components. Consequently, *Easy5* is particularly well suited for the highly transitory analysis required for hydraulic system modeling.

The pedigree of an analysis tool often indicates its primary strengths. Therefore, computer programs starting life as fluid systems modelers tend to have more capabilities related to pressure drop and flow predictions than competitor products, where fluid flow was a secondary focus. Mentor Graphics FloMaster [4] illustrates this point, adding the capability to generate pressure drop *loss coefficients* (*k-values*) (Section 9.3.3.2) using a simplified *component 3-D CFD* simulation for use in 1-D network models (which they call *system 1-D CFD*).

10.3.1 System 1-D CFD

Some software companies, including FloMaster, advertise offering 1-D CFD for their network analyzer. CFD uses numerical methods to predict airflow fields and their effects on temperatures, pressures, mass, and heat transfer in one, two, or *three-dimensional* (3-D) space (Section 10.5).

FloMaster combines 1-D *k-values* with a CFD equation solver to calculate flow rates and pressure drops. CFD solvers like the SIMPLE algorithm that ACCUEX uses (another 1-D CFD code) can arrive at converged solutions previously unreachable with traditional 1-D network flow modelers. This is especially true for highly transitory flow, like water hammer with its rapid pressure spike. These codes fail, however, to address the biggest source of errors for most ATM-related network modeling, user input *k-values* and nonuniform flow velocities that require 3-D CFD to capture.

10.3.2 Component 3-D CFD

Component 3-D CFD simulations replace the 1-D characterization with a 3-D mesh to calculate an equivalent *k-value* based on pressure drop calculations generated by CFD. It supports modeling the actual geometry contributing to the flow resistance in a duct or other component, instead of searching for published *k-values*. Data which is often unavailable for the identical configuration and desired flow conditions. Spatial constraints on aircraft force designers to deviate from common design geometries that published *k-values* are more likely to cover.

A 3-D CFD simulation, therefore, addresses the major limitation of the 1-D network flow solver, a dependence on published *k-values* generated for a limited number of ducting configurations.

FloMaster includes a 3-D CFD modeling capability to calculate *k-values* for use in the 1-D flow models, an extremely useful enhancement. The program minimizes the resources required to increase the accuracy of the solution by running a series of flow rates (one time) to calculate a table of *k-values* versus Reynolds numbers.

10.4 Multi-domain/Co-simulation

Multi-domain physical systems modeling software supports the simulation of the impact of different physical phenomena on system performance, allowing engineers to better replicate real-world operations. These programs typically include separate modules to simulate different systems, such as hydraulics, electrical, ECS, and flight controls. They support *multidisciplinary design optimization* (MDO) in which the simultaneous design of all (systems) disciplines drive an optimum aircraft design. This captures the interactions between different systems during many simulations as system design parameters are varied to create a more efficient vehicle-level design.

While co-simulation refers to the ability to solve multiple domains simultaneously, an alternative approach involves calculating a converged solution for each model in succession before moving on to the next one.

The main limitation with today's multi-domain/co-simulation software is computing hardware and the time required to characterize the many complex systems on a large commercial airplane.

Popular 1-D multi-domain simulation tools include Mathworks *Simulink* [14] and Siemens *Amesim* [15].

10.5 Computational Fluid Dynamics

CFD involves solving the Navier-Stokes differential equations for the conservation of mass, momentum, and energy for a series of adjacent volumes. It is used to calculate pressure drops, heat transfer rates, design mass flow rates, and fluid dynamic forces such as lift and drag.

The first aircraft CFD models predicted airplane *external flows* [2] to help minimize airplane drag and maximize lift. The Douglas Aircraft Company (part of Boeing today) built a rudimentary 3-D method for analyzing fluid flow over an airfoil in 1967 [5].

By the late 1980s, CFD went beyond predicting the lift and drag characteristics of an air foil (wing) to include external and *internal flows* related to ATM.

10.5.1 External Flows

External flow modeling related to ATM provides in-flight aircraft skin pressure distributions where air enters or exits the vehicle external contours. This includes ram-air cooling system inlets and outlets, *overboard exhaust valves* (OEVs), and openings (drainage and ventilation holes) in fairings covering unpressurized volumes (*main landing gear* (MLG), wheel well, *air-conditioning* (AC) pack bay, leading and trailing edge, etc.).

10.5.2 Internal Flows

Internal flows correspond to fluid flow within the airplane skin, including systems and components covering

- Cabin airflow patterns, which determine passenger thermal comfort (Section 2.3.1) and convective heat transfer on contacting surfaces
- Air temperature gradients in cargo compartments
- Orifice requirements to balance flows in air distribution and *electrical/electronic* (EE) cooling systems
- In-flight ventilation flowrates through unpressurized volumes
- Thermal mixing of hot and cold air streams (which occurs in the mix manifold and trim air injection sites)
- Flow through systems components such as ducts, valves, and electronic boxes

10.5.3 Advantages and Limitations

The higher node density of CFD models compared to thermal and 1-D network flow models generates more accurate temperature and flow predictions when appropriately used. CFD can also produce impressive looking erroneous predictions in the hands of the unskilled analyst.

CFD can increase the accuracy of temperature predictions, replacing *convective heat transfer* (*h-value*) coefficients (Section 9.2.1.2), which use crude fluid velocity assumptions, with heat transfer predictions based on calculated fluid velocities.

A full 3-CFD model can dramatically reduce the errors in flow calculations common to a 1-D network flow simulation for a large ECS, like an air distribution system. Perhaps even replacing the need to build full-scale mockups to test proposed designs. This is beyond the capabilities of the latest FlowMaster 3-D component modeler.

Designers can verify and optimize that component designs involving complex 3-D flow patterns occurring from cabin overhead nozzles, trim air injectors, outflow valves, and ram are system inlets and outlets.

The major limitations to CFD applications are the much higher hardware and software requirements relative to a thermal/fluid system or network flow simulation program. How much higher? A 20,000-node SINDA thermal model of the entire external structure of a large commercial airplane could take 20 million thermal and fluid nodes to characterize with CFD. Therefore, while a typical thermal model requires a *central processing unit* (CPU) or two to generate temperature predictions within an acceptable time, the equivalent CFD mode can take 200 CPUs or more. Consequently, CFD is judiciously used in environments with limited computing resources.

10.5.4 Expanded Use

Much cheaper and faster multi-CPU workstations and servers, licensing arrangements that encourage *parallel processing* and reduced pre- and post-processing time are supporting a rapid expansion of CFD modeling.

Separate CPUs simultaneously solve equations for different portions of a model in parallel processing. This can dramatically speed up a model solution for the price of more computing hardware and software licenses.

Some CFD suppliers offer *parallel licenses*, which allow a model to parallel process among multiple CPUs if a *serial license* is purchased to run the simulation. As multi-CPU machine use expanded, CFD licensing agreements changed to encourage increased usage. Software suppliers reduced parallel license costs, contributing to increased penetration for analytical modeling previously dominated by thermal/fluid systems and 1-D network flow models.

Unlike thermal analysis and 1-D network flow codes, a plethora of CFD codes are available in the marketplace as the best and brightest in universities across the globe keep coming up with new ideas [6]. A few of the more widely used commercial codes in this dynamic field include AcuSolve [7], ANSYS Fluent [8], CFD++ [9], and Star-CD/Star-CCM [10].

While each of the major CFD codes has their strengths and weaknesses, most do a good job of simulating fluid flow and heat transfer with high accuracy. Often the deciding factors for choosing a code come down to licensing costs, and the time required to build the models (*pre-processing*) and prepare the data for viewing (*post-processing*). Faster and cheaper processors and cheaper parallel licenses have reduced simulation run times, making pre- and post-processing speeds an increasingly greater concern.

10.6 Pre-processing

Pre-processing involves generating the mathematical model used to calculate the desired parameters from an electronic graphics-based solid model built with a CAD program.

Similar to parallel processing, graphically based pre- and post-processing began with CFD but now includes thermal analysis codes.

The CAD solid models, which generate photographic-like visual representations of aircraft parts and systems, contain the geometry of structure and systems requiring temperature predictions, usually in excruciating detail, leading to an excessive number of parts that an analyst building a CFD or thermal model must import and repair. Some CAD model data may not be convertible to a form supporting CFD or thermal meshing, which model builders must delete.

Meshing, the next step in pre-processing, involves partitioning the solid model into a series of nodes sized to ensure an efficient solution. The goal in meshing is to identify the minimum number of nodes required to generate solid or gaseous predictions at an acceptable accuracy level and runtime. Eliminating parts of the CAD solid model having an insignificant effect on the model predictions and combining other parts are primary tasks in reducing the number of nodes. These are two very tedious and time-consuming processes. The model is then broken up into multiple nodes, which are solved for temperatures, pressures, and/or velocities.

While CFD and network thermal analysis tools include basic pre- and post-processing tools, alternative programs (with enhanced capabilities) are also available, which provide significant time savings. CFD pre-processing tools include SpaceClaim, ANSYS Preprocessor [11], and Pointwise [12].

The ANSYS Corporation markets SpaceClaim, in addition to producing the Fluent CFD code, illustrating the trend of larger corporations buying up the smaller companies producing supporting programs. SpaceClaim also produces a dedicated version for Thermal Desktop [13]. CFD users of Thermal Desktop, who were familiar with the reduced time required to build their CFD models, drove C&R Technology to add this needed capability.

10.7 Post-processing

Post-processing involves converting the numerically generated predictions into a form providing the required understandings. This usually includes a graphical view of all temperatures and/or flow parameters combined with numbers for select locations. Most models built today have way too many nodes to present more than a tiny percentage of the data as numerical values. Meanwhile, identifying a specific temperature from a rainbow color plot is fraught with errors, making a pure graphical plot usually inadequate for documentation.

Similar to pre-processing, post-processing tools with capabilities beyond those available with the codes generating the predictions are also available. Once again, ANSYS leads the way in offering CFD-Post as a complement to their Fluent CFD code, at an additional cost (of course).

10.8 General Programming Environments

Modeling many of the systems impacting ATM (mostly ECS) from the state equations using various programming environments and languages is another viable option. Thirty years ago FORTRAN was the dominant programming language for solving engineering problems and modeling physical systems. Today the market includes multiple programs such as MATLAB and Modelica. While this category includes the spreadsheet Excel, due to its widespread use in the engineering community, it comes with added risk and frustrations [16].

The Excel spreadsheets on most PCs offer a potentially convenient means of quickly modeling many ATM processes. Unfortunately, it comes with the added risk of generating faulty numbers, and the constant frustration of dealing with a program which becomes increasingly difficult and time consuming to use with each new version. I have run into the issue of Excel getting simple math wrong with increasing frequency during a 30-year career involving extensive analytical modeling of aircraft and spacecraft systems. Then there is the increasing number of steps required to complete the most common tasks with each upgrade.

10.9 Tool Source

Equipment and vehicle manufacturers can develop analytical tools in-house or purchase *commercial off-the-shelf* (COTS) software. There is even free software, called *open source*, which educational and non-profit organizations often use. *Governments* also offer programs (they write or finance) for free, or at a lower licensing cost, compared to commercial alternatives.

10.9.1 In-house

Engineers responsible for design or analysis often develop software to support their work. It can be as simple as a Microsoft Excel spreadsheet simulating an EE cooling system, or complex as a CFD analyzer developed 30 years ago, when COTS software options were more limited.

10.9.2 In-house versus COTS Software

In-house developed codes can provide multiple competitive advantages compared to COTS, even with dramatic improvements in COTS program capabilities over the past few decades.

In-house codes may more accurately model systems physics after years of development using test data software development houses lack. Automation of repetitive steps used in the analysis process is a second driving factor for in-house code development. While today's latest COTS software has significantly more automation capabilities (compared to past generations), it may still may not meet all desires of the aircraft manufacturer.

In-house codes also eliminate the cost of purchasing additional licenses to support expanded use. This includes maintenance contracts for upgrades and user support, which are typically 15 to 20% of the initial license price. Providing in-house support and maintenance and upgrades for a previously developed code may be cheaper than purchasing hundreds of additional licenses.

10.9.3 Open Source

Many *open-source software* (OSS) CFD packages, developed by universities or consortia, are available for free. Open-source software is open to users altering, improving, and distributing it. Some OSS prohibits commercial use, so legal departments must carefully review the fine print on licensing agreements before allowing downloads to company computers.

One of the most widely used CFD OSS programs called OpenFOAM [17] was originally created by Henry Weller in the 1980s at Imperial College, London. A not-for-profit and commercially backed foundation freely distributes two versions of OpenFOAM. A company, which makes a profit by offering technical support and an improved user interface, regularly updates the commercial version of OpenFOAM.

While the numerical performance of some OSS is comparable to the latest COTS alternatives, they often lack the polished user interfaces, user support, and regularly scheduled updates. Commercial companies may hesitate using OSS because of concerns over the lack of technical support, reliability, security, and legal issues.

Open-source spreadsheets, like Apache OpenOffice OpenCalc offer impressive, although reduced capabilities compared to Microsoft Excel, but none of the bloat or complexity, and much greater speed. Programmers contributing to open-source software seem to understand that doubling the number of steps required to complete a task does not improve software.

10.9.4 Government

Governments create or fund the creation of software aerospace manufacturers can or must use for contractual work on government programs. The *Fire Dynamics Simulator* (FDS), a CFD code developed by the US *National Institute of Standards and Technology* (NIST), is an example of a free government code applicable for airplane design [18].

NIST wrote FDS to support the development of smoke and fire detection and sprinkler systems for the building industry.

Analytical tool selection may also be contractually driven, with NASA long mandating that all suppliers use a version of the SINDA computer program.

10.10 Software Evaluation and Selection

Software evaluation includes choosing the type of software used to model a physical phenomenon and the analytical tool within the software category. Analytical tool evaluations consider their *efficiency*, *reliability*, *accuracy*, *compatibility*, and *cost*.

Efficiency includes the user effort and expertise and computing resources needed to achieve an acceptable solution. Reliability relates to generating a consistently accurate answer for a range of tool applications. Accuracy addresses differences between predictions and the in-service scenario analyzed, which includes that ability to simulate the phenomenon of interest. Compatibility relates to the ability to run models built using different software packages or support MDO. Cost includes the initial license purchase price, support contracts, and training.

Over time, the relative importance of each factor can change, although it usually boils down to economics (after meeting basic performance concerns). Thirty years ago, computing hardware costs (solver efficiency) was the primary concern. Today, however, the required head count to complete the entire analysis process may rank higher in importance since CPU costs have plummeted, courtesy of Moore's Law (Section 8.10), while labor rates continually rise. The multiple layers of managers, project managers, cost accountants, human resources, and *information technology* (IT) support personnel produce labor rates far exceeding the salary and benefits directly related to the analyst.

References

1. C&R Technologies, SINDA/FLUINT, accessed June 20, 2017, https://www.crtech.com/products/sindafluint.

2. MSC Software, "Sinda Advanced Thermal Simulation," accessed June 20, 2017, http://www.mscsoftware.com/product/sinda.

3. MSC Software, "Easy5 Advanced Simulation and Control Software," accessed June 22, 2017, http://www.mscsoftware.com/product/easy5.

4. Mentor Graphics Corporation, "FloMASTER Fluid Thinking for Systems Engineers," accessed June 22, 2017, https://www.mentor.com/products/mechanical/flomaster/flomaster/.

5. Hess, J.L. and Smith, A.M.O., "Calculation of Potential Flow about Arbitrary Bodies," *Progress in Aerospace Sciences* 8 (1967): 1–138.

6. CFD Online, "Codes," accessed August 3, 2017, https://www.cfd-online.com/Wiki/Codes#Commercial_codes.

7. Altair HyperWorks, "AcuSolve Overview," accessed June 22, 2017, http://www.altairhyperworks.com/product/AcuSolve.

8. ANSYS, "Ansys Fluent," accessed June 22, 2017, http://www.ansys.com/Products/Fluids/ANSYS-Fluent.

9. Metacomp Technologies, "Accurate CFD for All Regimes," accessed June 22, 2017, http://www.metacomptech.com/index.php/features/icfd.

10. Siemens, "Star-CD," accessed June 22, 2017, https://mdx.plm.automation.siemens.com/star-cd.

11. ANSYS, "ANSYS SpaceClaim," accessed June 22, 2017, http://www.spaceclaim.com/en/default.aspx.

12. Pointwise, "Pointwise the Choice for CFD Meshing," accessed June 2, 2017, https://www.pointwise.com/.

13. BETA Simulation Solutions, "ANSA Preprocessor," accessed June 22, 2017, https://www.beta-cae.com/ansa.htm.

14. MathWorks, "Simulation and Model-Based Design," accessed June 2, 2017, https://www.mathworks.com/products/simulink.html.

15. Siemens, "Optimize System Performance from Early Design Stages," accessed June 2, 2017, https://www.plm.automation.siemens.com/global/en/products/simcenter/simcenter-amesim.html.

16. Mathworks, "Matlab," accessed June 2, 2017, https://www.mathworks.com/products/matlab.html.

17. Wikipedia, "OpenFOAM," 2017, accessed September 29, 2017, https://en.wikipedia.org/w/index.php?title=OpenFOAM&oldid=793738858.

18. Yu, Z.J., Qiang, X., and Han, J.W., "The Application of FDS Used in the Cabin Fire Simulation and Human Evacuation of Civil Aviation," *Proceedings of 2012 International Conference on Mechanical Engineering and Material Science (MEMS 2012)*, November 2012, https://www.atlantis-press.com/proceedings/mems-12/3900.

11

Testing

Testing leads to failure, and failure leads to understanding.

—Burt Rutan [retired US aerospace engineer, designed the first plane to circle the globe without refueling]

11.1 Introduction

Thermal-related testing on an airplane program can characterize material thermal properties, validate analytical models, demonstrate aircraft system performance, and support government certification requirements to ensure passenger safety. Testing can also show compliance to the stringent requirements aircraft face from their hazardous operating environment.

This chapter includes guidelines for *identifying the need* for thermal-related testing and examples of different types of testing and instrumentation used in a laboratory setting and on an aircraft.

11.2 Identifying the Need

Testing to support *aircraft thermal management* (ATM) can be as simple as sending an 8-inch diameter coupon (sample) from the latest composite material system to a laboratory to measure *thermal conductivity* (k). It can also be as complex as flying an aircraft and crew to Siberia to demonstrate vehicle performance in a -40°F (-40°C) *outside ambient temperature* (OAT). The more elaborate and expensive tests require more evaluation to determine the absolute necessity. *Uncertainty* in the analysis, *certification requirements*, the *criticality* of passenger safety for the items tested, and *cost* are considered when developing an aircraft test plan.

11.2.1 Analytical Uncertainty and Design Margins

A history of accurate predictions shown on similar systems or structure (using a similar modeling approach) may provide sufficient technical justification for trusting analytical predictions with minimal or no testing.

Design margins also determine the need for testing. The structures group may forgo testing for designs predicted to never approach their minimum or maximum allowable load capabilities. Large structural design margins, however, are less likely with each successive airplane program as aircraft designers work to maintain the downward trend in fuel burn by eliminating unneeded structural material (weight).

11.2.2 Certification Requirements

A commercial aircraft must meet volumes of requirements covering the design, manufacture, build, and operation of the smallest part to the entire assembled airplane before government regulators allow them to carry passengers. These written requirements are open to interpretation, like a legal document presented in a court of law. Government regulators and company-employed designated technical experts can even sound like lawyers when discussing the intent of each individual requirement and the technical data needed to show compliance.

Technical data is the analytical or test-based information proving the airplane is safe for the operating conditions that can occur over the life of the worldwide fleet. The mix of analysis and test data providing the *means of compliance* is determined in discussions between the *Federal Aviation Administration* (FAA) certification office, in the US, and privately employed certification personnel.

The FAA authorizes private citizens, with a demonstrated expertise in a specific area, as *Designated Engineering Representatives* (DERs) and *Authorized Representatives* (ARs), to approve, or recommend approval, of technical data presented to the FAA. Employed by private companies as direct hires, or independent (self-employed) consultants, DERs and ARs save US taxpayers the cost of funding more government jobs, while taking advantage of the greater knowledge base of industry *subject matter experts* (SMEs). An SME is a person recognized as an authority in a specific area. While some FAA regulators are extremely knowledgeable about the technical details of the systems or structure they oversee, most lack the depth of understanding of an industry SME. Senior SMEs with years of experience working for an airframe manufacturer can identify potential safety issues an outside entity would miss. This is one reason airplane manufacturer internal safety requirements can exceed government regulations.

While having a company self-regulate provides a potential conflict of interest, having a less knowledgeable government worker assume the entire responsibility could result in a less safe and more expensive aircraft. In addition, the FAA will remove the DER/AR designation (pull their ticket) for individuals they believe are negligent in their duty to put passenger safety above the interests of the manufacturer. DERs/ARs take that risk very seriously, besides concerns about passenger safety, in their actions and decisions.

Europe and a few other countries, including Canada and Russia, have their own regulatory agencies which may increase certification requirements before allowing passenger service in their airspace. For example, Transport Canada and Russia are more stringent about cold-day operations than the FAA for obvious reasons. Most countries, however, accept FAA certification.

11.2.3 System Criticality

Items directly related to maintaining controlled flight, such as systems required to move a flight control surface, demand rigorous testing regardless of confidence in the analysis. Less critical systems, such as a lavatory/galley exhaust system, could forgo testing with adequate confidence in the analysis method.

11.3 Material Thermal and Surface Optical Properties

The most frequently used thermal property data on an airplane program are *thermal conductivity* (*k*) and *specific heat* (c_p) for materials, and *emissivity* (ε), *solar absorptivity* (α), and *solar reflectivity* (ρ) for surface coatings (paints). Windows add the additional requirement for *solar transmissivity* (τ) data, which glass manufacturers usually provide.

  The public domain and material suppliers are the first places to look for material property data. Proprietary materials developed by the aircraft and material manufacturers, which are more common with the composites, have increased the need for thermal testing. Structural materials under development undergo extensive testing programs to measure strength and strain behavior, unlike thermal properties, often forcing the thermal team to initiate testing.

11.3.1 Thermal Conductivity

Thermal conductivity is the ability of a material to conduct heat. Testing procedures for *k* are based on the material. Two commonly used tests are the *guarded hot plate* and *comparative cut bar.*

11.3.1.1 GUARDED HOT PLATE

The guarded hot plate uses a known heat source and two thermocouples to determine the thermal conductivity based on Fourier's law for one-dimensional conductive heat (Section 9.2.1.1).

$$k = \frac{q_{\text{heater}}\, l_{\text{sample}}}{A_{\text{sample}}\left(T_{\text{hot}} - T_{\text{cold}}\right)} \tag{11.1}$$

where
  k = thermal conductivity, Btu-in/h-ft^2-°F (W/m-K)
  q_{heater} = heater power, Btu/h (W)
  l_{sample} = test sample thickness, in (m)
  A_{sample} = test sample heat transfer area, ft^2 (m^2)
  T_{hot}, T_{cold} = material hot and cold side temperatures, °F (°C)

  Guarded refers to a heater around the edge of the sample that limits heat transfer by maintaining a zero-temperature difference. The guarded hot plate provides high-accuracy measurements for materials with stable and low *k*-values, like insulation. There are also unguarded hot plate test setups [1].

  High *k* materials, such as metals, produce insufficient temperature differences to overcome the error and measurement uncertainty using the guarded hot plate. The *comparative cut-bar method* test works well for those high *k* materials.

11.3.1.2 COMPARATIVE CUT-BAR METHOD

The comparative cut bar is based on the principle of comparing temperature gradients between samples of a known and unknown material.

A material with an unknown thermal conductivity is sandwiched between materials of known thermal conductivity. Heating one end of the layup while cooling the opposite end generates a heat flux. Calculating the heat flux using Fourier's law, based on the temperature drop through the known materials, provides the required input for calculating k for the unknown material.

11.3.1.3 COMPOSITE MATERIALS

Composite materials require k testing of samples cut both parallel and perpendicular to the primary fiber direction to capture the effect of fiber orientation. This can dramatically increase testing costs, compared to metals with their constant (isotropic) properties in all directions.

11.3.2 Specific Heat

Specific heat is the heat required to raise the temperature of a substance by a given amount. Knowing how quickly a material's temperature changes is required to predict transient temperature responses. Two typical transient temperature responses of interest are how long it takes to pull down the cabin air temperature for an unpowered airplane in a hot outside ambient environment, and how high the leading-edge structure temperature rises due to bleed air impingement from a failed duct seal.

For solid materials, c_p is measured by placing a heated sample into a liquid (water) bath of a known mass in an insulated container called a *calorimeter*. After the material and liquid temperatures equalize, the liquid temperature change indicates the heat transfer to the material required to change the material temperature from Equation (11.3).

$$Q_{bath} = \left\{ Mc_p \Delta T \right\}_{bath} \tag{11.2}$$

where

$\quad Q_{bath}$ = heat loss from liquid bath, Btu (J)
$\quad M$ = liquid bath mass, lbm (kg)
$\quad c_p$ = specific heat of liquid bath, Btu/lbm-°F (J/kg-°C)
$\quad dT$ = liquid bath temperature change, °F (°C)

$$c_{p_{sample}} = \frac{Q_{bath}}{\left\{ MdT \right\}_{sample}} \tag{11.3}$$

where

$\quad M$ = test material mass, lbm (kg)
$\quad dT$ = test material temperature change, °F (°C)

11.3.3 Infrared Emissivity and Reflectivity

Emissivity measures the ability of a material surface to emit electromagnetic radiation while reflectivity refers to reflecting energy from a surface. *Infrared* (IR) refers to the electromagnetic spectrum where heat transfer occurs.

Reflectance can be *specular*, like a mirror where the incident angle to the surface equals the reflected angle; *diffuse*, such as light striking asphalt pavement where the reflected energy occurs in all directions; or a combination of both.

Surfaces can also be opaque or transparent. For opaque surfaces, ε plus ρ equals one, so ρ measurements using an *infrared* (IR) *reflectometer* can determine ε. An IR reflectometer radiates pulses of IR light onto an object and measures the strength of the reflected IR light using a sensor called a *pyranometer*.

11.3.4 Solar Absorptance and Reflectance

Solar absorptance is the fraction of solar energy striking a surface absorbed and converted to heat. Calculated by looking at the amount of energy reflected, using a *spectrophotometer*, the ρ averaged over different wavelengths generates an α for use in analysis. The *American Society for Testing and Materials* (ASTM) *Standards E903* and *E892* document this method.

11.3.5 In-Service Degradation of Properties

11.3.5.1 MATERIALS

Thermal conductivity of some insulation materials can change following the absorption of liquid. This effect is greatest for the lightest insulation materials, where the loft retention is more vulnerable to decreasing. Since installations are designed to minimize water ingression, analytical predictions normally ignore this effect.

11.3.5.2 SURFACES

Most published aircraft surface coating thermal data corresponds to clean new surfaces, which differ from in-service performance values following dirt accumulation or *ultraviolet* (UV) radiation exposure. Increases in the ε of a hot metal duct or α of a white fuselage paint can have a large impact on structure temperatures.

Lacking industry standard methods for estimating in-service effects on material or surface thermal performance in aircraft applications, the responsible analysts must develop their own methods for accounting for it in their analytical models. This may include developing unique tests to estimate surface property ε or α changes.

11.4 FAA Fire Testing

The FAA has developed a variety of fire tests, over many years, to ensure passenger safety in the event of a fire occurring anywhere on an aircraft. Fire testing demonstrates the meeting of performance criteria by aircraft *materials* during flame impingement, or *systems* designed to *detect*, *suppress*, or *extinguish* a fire. An understanding of the fire testing assumptions is needed to simulate the same fires in analytical modeling of aircraft systems and structure.

11.4.1 Materials

Material fire testing evaluates the *burn resistance* of ignitable cabin materials (i.e., seat cushions), the ability to resist *flame penetration* for materials designed to contain a fire (i.e., cargo liners, engine fire walls), and *heat* and *smoke generation*.

11.4.1.1 FAA FIRE TEST HANDBOOK

The *FAA Fire Test Handbook* defines fire testing apparatus, procedures, and acceptance criteria for materials and systems designed to ensure aircraft survival and passenger safety in a fire event [2]. Fire testing equipment and procedures continuously evolve in response to aircraft fire events, making the *FAA Fire Test Handbook* a regularly updated working document.

11.4.1.2 **BURN RESISTANCE**

Burn resistance deals with the tendency of a material to self-extinguish after the removal of the fire source. A cabin full of highly burn-resistant materials can keep small manageable fires from becoming uncontrollable. The fire source for this testing ranges from a Bunsen burner to an oil burner.

Bunsen burner testing determines the resistance of materials to burning during a 12- or 60-s period. One performance criteria, a *drip flame time*, corresponds to how long a portion of the material may continue burning after falling from the specimen.

Other testing uses an oil burner to generate a much larger flame at a controlled temperature and heat flux, as defined in the *FAA Fire Test Handbook*.

Seat cushions, which provide a potentially large fuel source for a cabin fire, must pass a burn resistance test where flames generated by an oil burner engulf the entire item (Figure 11.1). The cushion weight loss must not exceed 10% following a 10-min flame exposure to pass the FAA acceptance criteria. The oil burner air intake is adjusted to provide a heat flux of 10 ± 0.5 Btu/ft²-s, (11.9 ± 0.6 W/cm²) or more, and a flame temperature no less than 1800°F (982°C).

FIGURE 11.1   Cargo liner and seat cushion burner tests [3].

© Federal Aviation Administration

11.4.1.3 **BURNTHROUGH**

The oil burner used for seat cushion testing is also used to demonstrate flame penetration resistance for materials isolating areas vulnerable to fires, such as the engine, *auxiliary power unit* (APU), and cargo compartments. High air temperatures measured adjacent to the material sample on the liner side opposite the flame indicate flame penetration.

Oil burner testing started in the 1950s with the powerplant components (engines and APUs). Samples of materials protecting powerplant components must contain a 2000°F (1093°C) minimum temperature flame with at least a 9.3 BTU/ft²-s (10.6 W/cm²) heat flux density for 15 min.

In 1984 the FAA added the same testing for cargo liner materials. During a 5-min *burnthrough* test, a 1600°F (871°C) or greater flame, producing a heat flux density of 7.5 Btu/ft²s (8.6 W/cm²) or greater, must not penetrate the cargo liner. Air temperatures measured adjacent to the liner side opposite the flame source, which are no greater than 400°F (204°C), indicate flame containment.

The burnthrough flame impingement exposure time accounts for the required performance of fire detection, suppression, and extinguishing systems in FAA required testing. The thermal team assesses the impacts of the thermal signatures identified in fire testing on aircraft designs. They usually use analytical models, often built with *computational fluid dynamics* (CFD) codes (Section 10.5). This ensures the airplane meets performance requirements designed to support *continued safe flight and landing* (CSFL) following a survivable fire.

11.4.2 **Systems**

Cargo compartments, engines, and APU fire detection, suppression, or extinguishing systems must meet performance requirements consistent with the burnthrough requirements for materials isolating the fire.

11.4.2.1 **CARGO**

Cargo compartment fire protection systems must meet *minimum performance standards* (MPS) for the proposed detectors and suppression system design [4]. The performance standards, which are based on testing using Halon 1301, must be met using alternative agents for new aircraft, since the developed countries banned Halon manufacturing in 1994 after scientific studies indicated that it depleted the earth's ozone layer [5].

The MPS defines test procedures to ensure that fire protection systems passing the cargo fire tests will suppress or extinguish a fire within the 5-min flame exposure time demonstrated during burnthrough testing.

Fire tests occur inside a simulated lower lobe cargo compartment of a wide-body aircraft that includes in-flight air leakage through door seals. Leakage is critical to the performance of fire suppressant systems, which must maintain a sufficiently low-oxygen content to prevent open flames from developing after the initial flame suppression. Too much air leakage can overwhelm the suppressant flow rate, allowing an uncontrolled fire. Cardboard boxes filled with shredded paper represent the cargo fuel source, while a heated wire ignites the paper in one box to start the test.

Three luggage configurations require testing, cargo stacked in containers, and in the open, in a bulk or palletized configuration. Bulk cargo consists of loose suitcases in a cargo compartment, as seen in narrow-bodied (smaller) aircraft. Loose luggage placed on pallets loaded outside the airplane is palletized. A net keeps the luggage from shifting on pallets. Dual aisle wide-body jets usually use luggage containers or pallets excluding the furthest aft section of the aft cargo compartment. The curvature of the fuselage near the airplane tail cone provides insufficient floor space for pallets or containers.

Cargo fires and fire detection and suppression systems can perform differently for each cargo loading configuration, so they require separate tests or analysis to demonstrate adequate performance.

Fires starting in cargo containers can grow for a longer time before detection by the smoke or fire detectors with the container walls impeding the flow of heat and smoke. After detection, the suppressant agent may take longer to penetrate the cargo container to reach the flames, allowing the fire to continue growing more rapidly compared to fires starting outside the container. On the other hand, the cargo container walls can reduce access to make up air to replace the consumed oxygen, reducing the fire growth. The container surfaces protect the cargo liners from flame impingement, the primary concern for a cargo fire. While radiation from a hot cargo container surface can provide a high thermal load on a cargo liner, it pales in comparison to convective heating from a 1600°F (871°C) (or higher) flame.

Acceptance criteria for passing the cargo fire detection and suppressant test is the temperature versus time curve for multiple thermocouples located in the simulated cargo compartment. The area underneath the curve must be below a specified value.

11.4.2.2 HEAT GENERATION

Heat generated by an aircraft fire can kill passengers, disable equipment, and damage structure. Aircraft materials that release less heat while burning, therefore, improve passenger safety. A *heat release rate* (HRR) defines the amount of heat generated by burning a material, which the FAA limits for materials used in passenger cabins.

The temperature increase of airflow used to cool a burning material sample placed in an insulated container defines the HRR. Radiant (IR) heaters raise the material sample above its autoignition temperature causing it to burn. Subtracting the energy produced by the heaters from the energy gained by the cooling provides the HRR.

11.5 Model Validation

The primary purpose of ATM thermal testing on an airplane program is to *validate* analytical models used for design and/or certification. *Validation* means demonstrating that the analytical model agrees with testing within the required accuracy. Temperature is the most frequently recorded thermal test data since that is the primary parameter most thermal models predict. Fluid flow and pressure drops come in a close second with systems modeling.

11.5.1 Temperatures

A wide variety of temperature measuring devices are available for testing including *thermometers*, *probes*, and *non-contact* devices.

11.5.1.1 THERMOMETERS

Temperature changes In a thermometer cause the fluid in a glass bulb to expand or contract, raising and lowering the fluid level in an attached tube, which is correlated to temperature.

The glass tube thermometer accuracy depends on the tube length, with a very tall tube covering a narrow temperature range providing higher accuracy. Poor portability and no means for automatically recording data typically limit their use to viewing the OAT at the crew shelters used during ground airplane testing,

11.5.1.2 PROBES

Resistance elements, *thermocouples*, and *semiconductors* are examples of probes, which are the primary means of measuring temperatures in aircraft related testing.

Resistance Elements. The resistance element *thermistor* uses a material (ceramic and polymer) exhibiting very large changes in electrical resistance to measure temperature changes. A small electrical current is sent through the thermistor while the measured resistance indicates a temperature.

A *resistance temperature detector* (*RTD*) operates identically to a thermistor, using a pure metal material to improve the accuracy [6].

Thermocouples. Dissimilar metals joined at a junction generate a temperature dependent current in a thermocouple. The thermocouple is the industry standard measurement device due to extremely low cost, durability, and a wide range of useful temperatures.

Non-Contact. Non-contact devices sense the IR (radiative emission) from a body. They include single reading, and camera field devices providing a colorized picture of temperature variations. Both provide instantaneous surface temperature measurements, unlike a contacting probe which requires a significant time to allow the tip temperature to stabilize. The camera view area measurements support more thorough model validation.

The major disadvantage of the non-contact IR sensors is potentially lower accuracy compared to probe temperature measurements. IR temperature measurements require an ε assumption. Therefore, errors in estimating the ε or variations in ε over a surface can cause large measurement errors. Although, calibrating the IR camera ε settings using a temperature probe may minimize these errors. Camera ε settings are varied until the IR measured temperature matches the value measure with a surface contact probe.

11.5.2 Fluid Flow and Pressure Drops

Hydraulics and ECS analytical models simulating flow in pipes and ducts and in open areas (like a passenger cabin) use measured flow rates/velocities during testing for validation. Common flow measuring devices include *pitot tubes*, *manometers*, *mechanical* and *electrical pressure sensors*, *hot-wire anemometers*, and *ultrasonic flow meters*.

Pitot tubes, manometers, and pressure sensors measure the total pressures rise of flow in a duct, which is converted to a velocity using Bernoulli's equation (Section 9.3.1).

11.5.2.1 PITOT TUBE

A pitot tube is a simple device for measuring the static, total, and velocity pressure head of a single point in an airstream (Figure 11.2).

FIGURE 11.2   Pitot tube schematic [7].

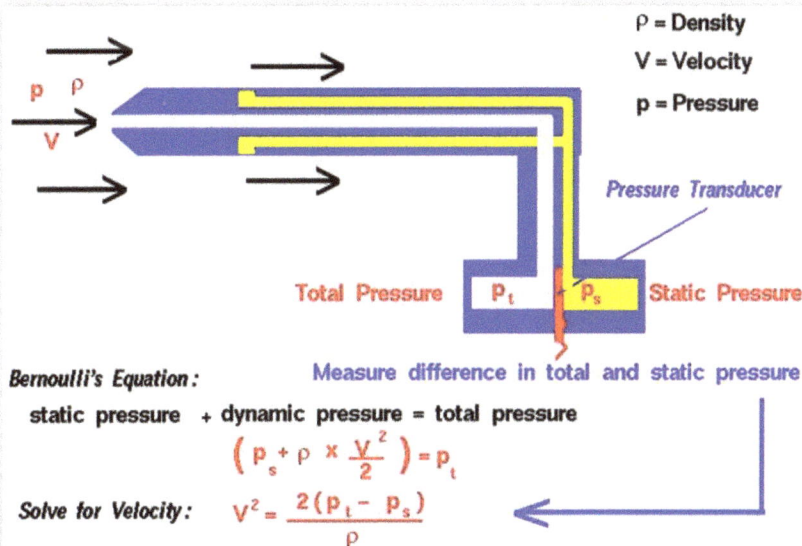

© NASA

A common pitot tube configuration consists of two tubes in the flow stream parallel to the flow direction with openings designed to capture the total and static pressure. The tube with the opening facing the flow (impact tube) captures the total pressure, while the tube with the opening perpendicular to the flow (static tube) captures the static energy, missing the impact of airflow impingement. You have felt the difference between the static and total pressure of moving air measured by a pitot probe if you ever held your hand out the open window of a moving car. The force felt by the hand is the total pressure while the background static pressure on your remaining body (shielded by the car windshield) goes unnoticed.

Pitot tubes are connected to a manometer, a mechanical or electrical pressure sensor, which displays the total pressure rise for conversion to a velocity. Aircraft depend on pitot probes for extremely reliable and accurate speed measurements.

11.5.2.2 **MANOMETERS**

Manometers, which use a column of liquid to indicate a velocity head (height), are simple, reliable (with no moving parts to break down), accurate, and inexpensive, just like pitot probes.

11.5.2.3 **MECHANICAL PRESSURE GAUGES**

Mechanical pressure gauges use mechanical elements, such as plates, shells, or tubes, which deflect during application of a pressure. This physical movement generates an electrical or mechanical output, such as a dial pointer moving through a series of linkages (Figure 11.3).

The mechanical approach of coupling a sensing element to the readout can introduce errors from the linkages, and a reduced frequency response due to the mass of the mechanical elements.

FIGURE 11.3   Pressure gauge [8].

© CEphoto, Uwe Aranas

11.5.2.4 **ELECTROMECHANICAL**

Electromechanical pressure sensors convert a pressure difference to an electronic signal, providing measurements for recording by a data-logging system. Bicking [9] describes multiple available electronic pressure sensors.

11.5.2.5 **HOT-WIRE ANEMOMETERS**

Hot-wire anemometers measure fluid velocities based on the cooling rate of a heated wire placed in a flow stream. The current in the wire is either varied to maintain a *constant temperature* or stays the same to maintain a *constant current*. An air velocity is, therefore, correlated to a current for constant temperature and temperature for constant-current hot-wire anemometers.

Constant-temperature anemometers are more common than constant-current anemometers due to a reduced sensitivity to flow variations.

The accuracy of hot-wire anemometers suffers when turbulent flow causes air to strike the wire from multiple directions. Placing the wire in a straight duct section, preferably no closer than 10 diameters from bends or transitions, minimizes this problem. Three hot-wire anemometers placed perpendicular to each other can pick up the velocity for less well-behaved airflow patterns outside a duct.

Hot-wire anemometer wires are very fragile and need frequent recalibration to maintain accurate measurements.

11.5.2.6 ULTRASONIC METERS

Ultrasonic flow meters measure fluid velocity using sound waves, which change frequency when affected by fluid motion. They are less commonly used than hot-wire anemometers due to higher costs and greater installation challenges.

11.6 Testing Boundary Conditions

Boundary conditions include the outside ambient environment, airplane and external heat loads, and aircraft performance parameters. The OAT, solar loading, wind speed and direction (for ground operations), and airplane Mach number, altitude, and *total air temperature* (*TAT*) (during flight) are typical outside ambient conditions collected for all aircraft testing.

Aircraft performance parameters for a *center wing* fuel *tank* (CWT) temperature test would include items such as fuel load and *air-conditioning* (AC) pack component internal air and external surface temperatures. Many system internal air temperatures are available as part of their built-in health monitoring, thus reducing the need for dedicated temperature measurements. Aircraft equipment manufacturers add instrumentation to identify reductions in performance, which point to impending component or system failure. This information is part of a growing field of airplane performance analytics used to prevent costly unexpected delays by proactively identifying and replacing parts that are likely to fail.

While surface and air temperatures and flow measurements dominate thermal testing, *relative humidity* (*RH*) measurements are also needed to better estimate sky temperatures and assess AC pack performance.

11.7 Testing Standards and Procedures

Equipment must pass a long list of thermal and other tests before receiving approval for use on an airplane. Testing starts with individual computer chips and microprocessors, and continues to the most complex airplane systems, like the *cabin air-conditioning and temperature control system* (CACTCS). The thermal team ensures that testing represents a sufficient portion of the design operating environment to meet certification requirements or validate analytical models. This entails being familiar with the latest analysis and testing procedures.

Industry trade organizations, *engineering societies*, and the *military* publish conditions and procedures designed to simulate the impact of in-service aircraft operation during ground, and environmental chamber testing.

11.7.1 Industry Trade Organizations

The *Radio Technical Commission for Aeronautics* (RTCA) is one of the most important industry trade organizations for defining *minimum* electronic thermal requirements to support testing. The term minimum is a reminder that some equipment may face more severe conditions that exceed their generic requirements.

Since 1958, the RTCA publication *Environmental Conditions and Test Procedures for Airborne Equipment*, *DO-160G* [10], or precursor, has been a primary standard for environmental qualification testing.

FAA Advisory Circulars reference *DO-160G*, making it a logical starting point for defining testing environmental conditions. *DO-160G* is, however, only a starting point, since it provides a small percentage of the information necessary to set up tests that accurately represent an in-service thermal environment for many installations. This is especially true for equipment outside the controlled ambient environment of the pressurized volume. For example, it proposes hot temperature testing for external equipment on the aircraft surface with no mention of solar heating, other than a 24-h diurnal temperature test. The one external solar loading graph is for the maximum horizontal flux (which does not mention the surface orientation), while equipment or lights located on vertical surfaces receive a different solar loading over time. There is also no mention of the need to insulate the back plate of lights or equipment mounted on an insulated fuselage skin.

11.7.2 Engineering Societies

Engineering societies, such as the *Society of Automotive Engineers* (SAE International), publish standards and *aerospace recommended practices* (ARPs) to help characterize the aircraft environment and testing procedures. The SAE Aerospace Division publishes multiple documents, such as *SAE ARP 986C, Guide for Qualification Testing of Aircraft Air Valves*, which provide useful guidelines for every aspect of testing from simulating the test setup to writing the test report.

11.7.3 Military

The US military also provides testing guidelines for equipment installed in military aircraft that commercial aircraft manufacturers can use. *Environmental Engineering Considerations and Laboratory Tests* (*MIL-STD-810*) defines test chamber environments and testing procedures that replicate the effect of in-service exposure.

Testing includes maximum and minimum survival temperatures, and lifetime temperature cycling using less extreme values. While the procedures and environments generally ensure that equipment installed on an on airplane is robust enough to meet in-service performance requirements, exceptions can occur.

11.8 Systems Testing

Much aircraft equipment operates as part of a system, where the performance of one component affects the performance of others. Examples include an AC pack with turbines, compressors, fans, and heat exchangers, each affecting each other.

Systems testing in a lab identifies operational issues prior to installation on the airplane. Since the testing is usually done at room temperature, component temperature data validates analysis methods used to simulate performance in extreme OAT conditions.

Lab testing often provides a great opportunity to gather supplier component data, such as temperatures of hydraulic fluid system actuators, or power feeders.

11.9 Airplane Testing

Airplane testing on ground and during flight ensures the safety, quality, and efficiency of the aircraft. While new airplanes undergo more testing, usually with multiple vehicles for large commercial airplane programs, it continues at a lower level for each airplane coming off the assembly line.

Reasons for airplane testing include *technology development* (to support a new airplane program or performance improvements), *certification* (to ensure passenger and crew safety), *nautical air miles fuel burn* testing (for a new airplane or following enhancements to reduce fuel consumption) and the *first flight* for every airplane built.

The thermal team is actively involved in airplane testing for new airplane programs, identifying minimum acceptable ambient environmental parameters for analytical model validation or certification testing. Flight test organizations can rarely find the hottest or coldest OATs covering airplane certification due to the extremely low probability of occurrence. Therefore, the thermal team may need to determine how much deviation is allowable while still demonstrating adequate performance by *data extrapolation* or *model validation*.

11.9.1 Data Extrapolation

If the OAT is too low for a hot-day or too high for a cold-day requirement, measured equipment, structure, or airplane air temperatures may be changed by an amount equal to the OAT shortfall to estimate whether the area of interest exceeds its design temperature.

$$T_{\text{extrapolated}} = T_{\text{measured}} + \left(T_{\text{OAT\_required}} - T_{\text{OAT\_measured}} \right) \tag{11.4}$$

where

$T_{\text{extrapolated}}$ = extrapolated temperature if OAT required for design or certification is not met
T_{measured} = measured temperature
$T_{\text{OAT\_required}} - T_{\text{OAT\_actual}}$ = difference between desired and actual boundary temperature

Since there is rarely a one-to-one correspondence between aircraft and outside ambient temperatures, analytical models are used to predict the actual correspondence when design temperatures fail to meet the requirements.

References

1. Schindler, A., Neumann, G., Stobitzer, D., and Vidi, S., "Accuracy of a Guarded Hot Plate (GHP) in the Temperature Range between −160°C and 700°C," *High Temperatures—High Pressures* 45, no. 2 (2016): 81–96.

2. Office of Aviation Research, *Aircraft Materials Fire Test Handbook*, DOT/FAA/AR-00/12 (Washington, DC: Office of Aviation Research, April 2000).

3. Ochs, R.I., "Fire Test Burner Methods," presentation at *FAA Safety Overview*, Singapore, February 7, 2012.

4. Reinhardt, J.W., "Minimum Performance Standard for Aircraft Cargo Compartment Halon Replacement Fire Suppression Systems (2nd Update)," DOT/FAA/AR-TN05/20 (Springfield, VA: U.S. Department of Transportation and Federal Aviation Administration, June 2005), https://www.fire.tc.faa.gov/pdf/tn05-20.pdf.

5. Bennett, R., "Replacing Halon in Fire Protection Systems: A Progress Report," *BoeingAERO*, QTR_04.11, April 2011, http://www.boeing.com/commercial/aeromagazine/articles/2011_q4/3/.

6. Wikipedia, "Thermistor," 2017, accessed May 20, 2017, https://en.wikipedia.org/w/index.php?title=Thermistor&oldid=774733509.

7. NASA, "Pitot-Static Tube," NASA Glenn Research Center, accessed June 3, 2017, https://www.grc.nasa.gov/www/k-12/airplane/pitot.html.

8. Wikipedia, "Pressure Measurement," 2017, accessed May 20, 2017, https://en.wikipedia.org/w/index.php?title=Pressure_measurement&oldid=779961537.

9. Bicking, R.E., "Fundamentals of Pressure Sensor," *Technology Sensors Online, 2018*, 2018, November 1, 1988, accessed May 20, 2017, http://www.sensorsmag.com/components/fundamentals-pressure-sensor-technology.

10. RTCA, Inc., "Environmental Conditions and Test Procedures for Airborne Equipment," Report No. DO-160G, December 8, 2010.

Military Aircraft Thermal Management

"Whoever said the pen is mightier than the sword obviously never encountered automatic weapons."

—Douglas MacArthur [WWII US General]

12.1 Introduction

Aircraft thermal management (ATM) on military aircraft shares numerous similarities with commercial aircraft. Both operate from similar ground ambient conditions based on a probability of occurrence. Both have *environmental control systems* (ECS) required to sustain life and ensure safe aircraft performance during hazardous external conditions (icing). Fixed-wing military and civilian aircraft both have flight-control surfaces, landing gears, and doors powered by hydraulically or electrically driven actuators, creating similar thermal issues. Both use the wing fuel as a primary heat sink. The *electrical/electronic* (EE) equipment temperature requirements even come from the same documents, in many cases, showing the similarity in thermal environments. With the extensive use of composites on the latest civilian aircraft, they even face similar issues with more temperature-sensitive structure.

While extensive similarities exist between ATM systems on commercial and military aircraft, additional mission requirements and newer technology increase the required analysis and testing.

A commercial aircraft has a *single mission* of ferrying passengers, which has remained consistent for the past 60 years. Military tactical aircraft, on the other hand, must meet *multiple missions* that continually evolve with technology, increasing the required analysis and testing. Systems allowing flight crews to locate and/or destroy an enemy before the same happens to them, or meet new mission needs such as in-flight refueling, add a new set of challenges beyond anything a commercial airplane sees. Variation in military aircraft, which range from military versions of commercial airplanes to supersonic fighters, adds thermal requirements lacking on commercial airplanes.

This chapter covers the unique ATM challenges associated with military derivatives of commercial aircraft and pure military designs.

12.2 **Commercial Airframes**

While fighter and bombers dominate media coverage of United States (US) aircraft programs, due to their massive cost overruns, the US *Department of Defense* (DOD) has identified dozens of mission types, which do not require the extreme maneuverability of an F-15 fighter or stealth of a B-2 bomber [1].

Military versions of commercial airframes successfully meet many of these missions at a significant cost savings to the taxpayer. The best-known missions include *Anti-Submarine Warfare* (ASW), battlefield and air traffic *Command and Control*, *Electronic Warfare*, *Reconnaissance*, air *Refueling*, and *Troop Transport*.

12.2.1 **Advantages**

In addition to expensive weaponry and countermeasures, airframe and software development costs drive the high price of the latest military aircraft. Small production runs to spread development costs create stratospheric prices for military vehicles, exceeding $1.7 billion (USD 2017) for each US-made B-2 bomber [2].

While a large commercial airplane program normally produces over 1000 airplanes during its life, the B-2 production only reached 20 airplanes due to high costs. A vicious cycle occurred with higher costs leading to fewer sales, leading to even higher per-unit costs.

Commercial sales pay the original airplane development costs for military derivatives. Turning a commercial aircraft, like the Boeing 767, into a tanker or *airborne warning and control system* (AWACS) vehicle (with its huge mounted radar) is, therefore, usually cheaper than building an airplane from scratch, even after significant structural upgrades. This is especially true for military capabilities requiring less extensive structural design changes, like a tanker.

Commercial airplanes can also fly a much higher percentage of the time, with less maintenance and quicker turnaround times between flights than most dedicated military airplanes. Meanwhile, the extensive worldwide parts network Boeing and Airbus maintain for commercial aircraft ensures a quick return to flight following expected system and component failures. Perhaps most importantly, commercial aircraft have proven to be very effective during many conflicts.

12.2.2 **Disadvantages**

The major disadvantages of using a commercial aircraft to accomplish military missions are possible reduced capabilities and a less optimized design.

While an impressive number of commercial airframes successfully serve military missions, slight changes in their operating envelope may have unexpected negative consequences. A classic example occurred on an E-6 aircraft, a Boeing 707. Following four decades in service, parts of the vertical tail fell off after the *United States Air Force* (USAF) changed its flight profile to meet new mission requirements [3].

Commercial aircraft may also lack the cooling capabilities necessary to handle the electronics required to act as air traffic control when launching hundreds of sorties (bombing runs) per hour, tracking enemy combatants, or jamming their radar. This brings up another issue with commercial airplanes. The navy gray paint designed to blend into the background is a poor substitute for the stealth of a modern dedicated military aircraft. Even with weapons jamming capability, the 767 is a big target with a large heat sink from a ram-air system hot-air exhaust, which may be absent from totally new designs. In addition, commercial aircraft designed to pack in the maximum number of passengers may be less efficient in handling military personnel and equipment.

12.2.3 Military Avionics

Even the most similar military version of a commercial aircraft, the dedicated troop transport, may need avionics upgrades to meet more demanding military requirements, such as providing more secure communications or navigation. Others are unique to military vehicles, including sensors and supporting electronics designed to identify and disable enemy threats (electronic countermeasures). The sensors may require more precise temperature control than a typical commercial aircraft electronic box, leading to cooling system redesign.

Further from commercial operations design are the airborne warning and command post aircraft with their massive amount of electronics and sophisticated radars. The Boeing-built 767 AWACS aircraft is the most recognizable example of this type of military derivative, with a rotating dome (rotodome) mounted above the fuselage (Figure 12.1).

FIGURE 12.1 Boeing 767 AWACS.

© Boeing

A hydraulic system rotating the rotodome generates heat, as does the radar it contains. The radar directs fighter aircraft to their targets hundreds of miles away, beyond the range of *surface-to-air missiles* (SAMs). Electronics and flight crew displays fill the main deck, which carries hundreds of passengers in commercial operations, while the lower lobe contains even more electronics [4].

While the AWACS is the Porsche of warning and command post aircraft, an air force with shallower pockets can buy the 737 *airborne early warning and control* (AEW&C) (Figure 12.2).

The 737 AEW&C contains a stationary radar mounted atop the fuselage, which requires a thermal design to ensure adequate heat rejection.

CHAPTER 12

FIGURE 12.2   737 AEW&C aircraft.

© Boeing

Commercial aircraft have even made inroads into the bomber arena with maritime reconnaissance provided by the 737 P-8 Poseidon *ASW*. The P-8 includes a bomb bay. Commercial aircraft are well suited for maritime reconnaissance since it requires the ability to spend long hours searching the oceans for a moving target at a typical commercial aircraft speed. They face thermal issues similar to other military derivatives with their large electronic equipment sets to cool.

12.3 **Bombers**

Bombers attacking well-defended land-based targets can depend on speed to reach the target and return home safely, or stealth to avoid detection.

Higher speed operation provides thermal challenges from the increased structure temperatures due to aerodynamic heating and a higher boundary air temperature to reject heat (Section 2.2.2). Stealth can mean the elimination of ram-air cooling to avoid the heat signature of the hot-air exhaust and thermal designs to hide the engine exhaust plume and thermal signature.

The latest USAF bomber, the B-2 Spirit (Figure 12.3), with its bat shape and radar-absorbing materials designed to minimize a radar signature, won the stealth versus speed contest, when Northrop Grumman won their proposal for a similar design for the next-generation bomber [5].

FIGURE 12.3   Grumman B-2 bomber.

© US Air Force

The B-2 has a similar cruise speed compared to a commercial jet (Mach 0.85) with a maximum speed capability of Mach 0.95. The Boeing B-52 bomber, which first flew in 1952, operates at a maximum speed of Mach 0.86 [6] (Figure 12.4).

FIGURE 12.4 Boeing B-52 bomber [6].

© Boeing

Developed in the late 1940s, the B-52 will probably still be flying decades from now, with electronics and engines upgraded multiple times over the past 60-plus years. A severely limited production run of B-2 bombers has kept the B-52 out of the scrap heap far longer than the original designers ever dreamed of. The B-52 bomber cost $90 million (2017 USD), nearly 20 times less than the B-2 in today's dollars [7].

Speed was the initial focus of the Rockwell B-1 Bomber (designed before the B-2). The prime contractor, Rockwell International, built three B-1 bombers with a top speed of Mach 2.2, before the US Congress cancelled the program in 1977 [8].

When production resumed on a different version of the B-1 in the early 1980s, it included new flight controls, upgraded avionics and electronic countermeasures, and stealth capabilities, typical upgrades for military aircraft. One surprising change was a reduced maximum speed of Mach 1.25, a rare instance in aircraft design when some thermal requirements became less demanding with a vehicle upgrade.

12.4 **Fighters**

Unlike bombers, for many decades all fighters have operated at supersonic speeds, which cause the nightmarish high ambient boundary temperatures for thermal designers (Section 4.3.2). Half a century ago, the Sukhoi Su-27, a twin-engine fighter plane built by the former USSR, reached a maximum speed of Mach 2.35, 1550 mph (2500 km/h) [9].

The Sukhoi was in response to the US-made F-111 Aardvark, which entered service in 1967 with a top speed of Mach 2.5, 1650 mph (2655 km/h) [9]. Fighters also include stealth features designed to reduce the aircraft thermal signature, such as eliminating ram-air-cooling systems. This means the fuel may need to absorb waste heat generated by *air-conditioning* (AC) packs, bleed air, electronic cooling, in addition to hydraulic systems. Hydraulic system and engine cooling are the only fuel heat sink users on most commercial airplanes (Section 5.4).

Greater reliance on the fuel heat sink for military aircraft creates higher fuel temperatures, which increases the flammability risk. Much costlier (compared to commercial aircraft) inerting systems (Section 8.9.2.2) remove more oxygen from their *nitrogen-enriched air* (NEA) stream, eliminating the possibility of the ullage igniting.

Fighter aircraft ECS also supply pressurized air to flight crew suits, in addition to cooling a tiny cockpit with a small AC pack.

12.5 Vertical Lift

Vertical-lift vehicles (helicopters) use rotating blades (rotors) pushing air downward to provide lift and forward motion after a minor tilting of the blades. The inefficiency of this process, compared to using a wing to provide lift, limits their operation to lower altitudes, where the air density is greater. The additional drag from higher air density also limits their airspeed. The AH-64E (Apache Guardian) attack helicopter's 21,000 ft (6.4 km) service ceiling and 186 mph (300 km/h) top speed is typical for vertical-lift aircraft [10].

While helicopters are great for taking off and landing from a small open field and hovering in place, they cannot compete with the slowest jet in traveling from point A to point B quickly and efficiently.

To get around this issue of speed and efficiency, the Bell and Boeing Vertol Helicopter companies began collaborating in the 1980s to develop the Boeing-Bell V-22 Osprey tilt rotor transport (Figure 12.5).

FIGURE 12.5   V-22 Osprey.

© Boeing

The V-22 Osprey combines the advantages of fixed-wing and vertical-lift aircraft using propellers to provide vertical lift during takeoff and landing from a constrained area, and a wing during flight. This allows it to fly higher and faster than pure vertical-lift helicopter, reaching a maximum speed of 311 mph (500 km/h) [11].

Critical thermal issues for helicopters include blade temperatures, heat generation in the gearbox (where engine movement translates into blade rotation), and avionics overheating due to a more constrained space and higher ambient temperatures.

Helicopter rotor blades, which are increasingly composite, face high solar loads with their black (high solar absorptivity) color and often nearly perpendicular orientation to the *solar noon* sun (Section 4.2.3). This can raise a blade temperature far above the *outside ambient temperature* (OAT) on ground with no movement. During flight, rotational tip velocities reaching up to 450 mph (200 m/s) drive the blades toward the *total air temperatures* (TAT), overwhelming the solar heating affect. The identical process occurs with a fixed-wing aircraft, with elevated skin temperatures (on ground from solar heating) approaching the *adiabatic wall temperature* (Section 2.2.2.3) shortly after takeoff.

Helicopters operate close to the ground for extended periods, where higher OATs make cooling of the avionics and cabin more challenging compared to a commercial transport cruising at 36,100 ft (11,000 m). The typical 787 or A350 spends more than 50% of a 24-hour day in flight, while airlines continually search for new ways to maximize time spent in flight-generating revenue.

The electronics bays are also more space constrained in helicopters than fixed-wing commercial airplanes, further contributing to the risk of overheating and equipment failures.

12.6 Directed Energy Weapons (DEWs)

A *directed energy weapon* (DEW) emits highly focused energy, often lasers (light), to damage or destroy a potential target. Weaponized lasers offer the potential for more precise control of damage compared to firing artillery rounds, with adjustments to destroy or disable different materials and avoid collateral damage [12]. Military strategists envision using lasers to disable vehicles or equipment in civilian areas, without killing innocent bystanders.

Recognizing their potential value, US military organizations, such as the *Defense Advanced Research Projects Agency* (DARPA), the Air Force, and Naval Research Laboratories, have poured millions of dollars into their development since the 1970s [13].

In response to government research, aircraft thermal engineers are investigating design options necessary to accommodate the large transient heat loads production DEWs are likely to produce on existing and future military vehicles. Raytheon even reported on June 26, 2017 that they had successfully hit an unmanned target with a laser mounted on a US Army Apache AH-64 helicopter.

References

1. Norris, G., "Boeing Reveals 737 MAX Military Derivative Study," *Aviation Week & Space Technology* (May 5, 2017).

2. US Air Force, "B-2 Spirit," December 16, 2015, http://www.af.mil/About-Us/Fact-Sheets/Display/Article/104482/b-2-spirit/.

3. Mutty, M.S., "A Comparison of Military and Commercial Aircraft Development," Report No. NDU_ICAF-93-F3, p. 23, http://www.dtic.mil/dtic/tr/fulltext/u2/a276830.pdf.

4. Nagabhushana, S. and Sudha, L.K., *Aircraft Instrumentation and Systems* (I. K. New Delhi: International Pvt. Ltd., 2010).

5. Seligman, L., "Northrop Grumman Wins Air Force's Long Range Strike Bomber Contract," *Defense News*, October 27, 2015, http://www.defensenews.com/story/defense/2015/10/27/northrop-grumman-wins-usaf-bomber-contract/74661394/.

6. Wikipedia, "Boeing B-52 Stratofortress," 2017, accessed October 8, 2017, https://en.wikipedia.org/w/index.php?title=Boeing_B-52_Stratofortress&oldid=808964880.

7. US Air Force, "B-52 Stratofortress," December 16, 2015, http://www.af.mil/About-Us/Fact-Sheets/Display/Article/104465/b-52-stratofortress/.

8. Military Factory, "Rockwell B-1 Lancer Long-Range Strategic Heavy Bomber Aircraft," accessed June 5, 2017, http://www.militaryfactory.com/aircraft/detail.asp?aircraft_id=27.

9. Chow, D. "Supersonic! The 11 Fasters Military Airplanes," *Live Science*, March 22, 2016, https://www.livescience.com/39829-fastest-military-airplanes.html.

10. Military Today, "AH-64E Apache Guardian," accessed June 10, 2017, http://www.military-today.com/helicopters/ah_64e.htm.

11. Boeing, "V-22 Osprey," Boeing/Defense/V22 Osprey Technical Specifications, accessed June 10, 2017, http://www.boeing.com/defense/v-22-osprey/.

12. Hawkins, D., "Laser-Equipped-Helicopter Zaps Its First Target, to Defense Contractor's Delight," *Washington Post*, June 27, 2017, https://www.washingtonpost.com/news/morning-mix/wp/2017/06/27/laser-equipped-helicopter-zaps-its-first-target-to-defense-contractors-delight/?utm_term=.001e63d88226.

13. Atherton, K.D., "What's Next for the F-35B? Lasers," *Popular Science*, August 30, 2016, http://www.popsci.com/whats-next-for-f-35b-lasers.

nomenclature

Introduction

The nomenclature summarized below includes units, Roman and Greek letters used as equation parameters, dimensionless parameters, subscripts, and acronyms. Each chapter includes a more thorough description at its first introduction.

Units

° - degrees
°C - degrees Celsius
°F - degrees Fahrenheit
°R - degrees Rankine
Btu - British thermal units
ft - feet
h - hours
J - joules
K - kelvin
kg - kilogram
km - kilometer
kPa - kilopascal
lbf - pound-force
lbm - pound-mass
m - meters
mi - miles
s - second
W - watt
km - kilometer
kW - kilowatt
cfm - cubic feet per minute
mph - miles per hour
MW - megawatt
in - inch

Roman Letter Symbols

A - Area for heat transfer
A_0 - Daily or yearly temperature fluctuation amplitude for ground temperature equation
AM - Air mass model optical length
A_{sample} - Material sample heat transfer area for thermal conductivity measurements
c - Speed of sound

C - Thermal capacitance

CO_2 - Carbon dioxide

c_p - Specific heat at constant pressure

d - Damping depth, the distance below the soil surface where soil temperature amplitude decreases to $1/e$ (1/2.78) of the soil surface A_0

D - Tube diameter

D_h - Hydraulic diameter

dT - Temperature difference or change

f - Frequency

f_{cl} - Fraction of clothed/nude surface area, used in thermal comfort calculations

f_D - Darcy-Weisbach friction factor

f_{D_lam} - Darcy-Weisbach friction factor for laminar flow

f_{D_turb} - Darcy-Weisbach friction factor for turbulent flow

f_f - Fanning friction factor

g - Acceleration due to gravity

g_c - Gravitational constant unit conversion factor

G_h - Convection thermal conductance

G_k - Conduction thermal conductance

G_r - Radiation thermal conductance

h - Convective heat transfer coefficient

H_2O - Water

Hg - Mercury

I - Electrical current

I_0 - Solar radiation received by earth's atmosphere

I_{cl} - Clothing thermal insulation used in thermal comfort calculations

I_{sc} - Solar constant

k - Thermal conductivity

K - Pressure loss coefficient

k_f - Thermal conductivity of a fluid

L - Length

L - Thermal load used in thermal comfort calculations

l_{sample} - Material sample heat flow length for thermal conductivity measurements

M - Metabolic heat generation used in thermal comfort calculations

$\dot{m}$ - Mass flow rate

Ma - Mach number

$m_{aircraft}$ - Aircraft mass

m_{brake} - Brake mass

$m_{heat\ sink}$ - Heat sink mass

N_2 - Nitrogen

O_2 - Oxygen

O_3 - Ozone

P - Perimeter

P - Power

p - Pressure

p_a - Partial pressure of water vapor

$p_{dynamic}$ - Dynamic pressure

p_s - Static pressure

p_{sat} - Saturated vapor pressure

p_t - Total pressure

p_{t1} - Upstream total pressure

p_{t2} - Downstream total pressure

p_v - Velocity pressure

q - Heat transfer rate

Q_{bath} - Heat loss from a liquid bath to a material sample during specific heat testing

q_{heater} - Heater input power

q_{in} - Heat transfer into a fluid volume

r - Recovery factor for pressure or temperature

R - Gas constant for air

r_1 - Cylinder inner radius

r_2 - Cylinder outer radius

R_{cl} - Clothing thermal resistance used in the thermal comfort equation

R_{elec} - Electrical resistance

RH - Relative humidity

R_h - Convection thermal resistance

R_k - Conduction thermal resistance

R_r - Radiation thermal resistance

R_{xxx} - The net resistance for multiple resistances in series or parallel

s - Entropy

$S_{reflected}$ - Solar load reflected from the ground on to a surface

S_{direct} - Solar load received directly

t - Time, hours since

t_0 - Time lag from an arbitrary starting point for the minimum temperature, *hrs* for diurnal and *days* for yearly ground temperature predictions

T - Temperature

$T(z,t)$ - Soil temperature at a depth z and time t

T_∞ - Bulk fluid temperature

T_a - Air temperature

T_{amb} - Ambient air temperature

T_{ave} - Daily or yearly average temperature of the ground temperature equation

T_{aw} - Adiabatic wall temperature

T_{bound} - Boundary temperature

T_{cl} - Clothing average temperature in the thermal comfort equation

T_{dp} - Dew point temperature

T_{equiv} - Equivalent temperature, refers to extrapolations of test data

$T_{extrapolated}$ - Extrapolated design temperature if *outside ambient temperature* (OAT) required for design or certification is not met

v - Velocity

Greek Letter Symbols

α - Altitude/elevation angle in solar load calculations

α - Solar absorptivity

α - Thermal coefficient of resistance of a thermocouple

α - Thermal diffusivity

$\Im$ - Shape or view factor for radiative heat transfer between surfaces

β - Angle of surface relative to the ground for solar load calculations

β - Volumetric thermal expansion coefficient

δ - Declination angle

Δ - Delta (difference)

ε - Emissivity

ε - Duct absolute surface roughness
ϕ - Latitude for solar load calculations
ϕ - Surface angle for Grashof number calculation
γ - Kinematic viscosity
γ - Ratio of specific heats
η - Efficiency
μ - Dynamic viscosity
ρ - Density
ρ - Solar reflectivity
σ - Stefan-Boltzmann constant
Σ - Summation
τ - Solar transmissivity
ω - Radial frequency for soil temperature equation
ω - Hour angle for solar flux calculations

Dimensionless Groupings

Gr - Grashof number
Ra - Rayleigh number
Re - Reynolds number
Pr - Prandtl number
Nu - Nusselt number

Acronyms

1-D - One Dimensional
3-D - Three Dimensional
1-G - One gravity
AC - Air Conditioning
AC - Alternating current
ACM - Air Cycle Machine
AEW&C - Airborne Early Warning and Control
ALCM - Air-Launched Cruise Missiles
AM - Air Mass Model Optical Length
APU - Auxiliary Power Unit
AR - Authorized Representative
ARP - Aerospace Recommended Practice
ASHRAE - American Society of Heating, Refrigerating, and Air-Conditioning Engineers
ASN - Aviation Safety Network
ASTM - American Society for Testing and Materials
ASW - Anti-Submarine Warfare
ATM - Aircraft Thermal Management
AWACS - Airborne Warning and Control System
BOL - Beginning of Life
BTMS - Brake Temperature Monitor Sensor
CAC - Cabin Air Compressor
CACTCS - Cabin Air-Conditioning and Temperature Control System
CFD - Computational Fluid Dynamics

CFRP - Carbon Fiber Reinforced Plastic
CO$_2$ - Carbon Dioxide
COTS - Commercial Off-the-Shelf
CPI - Consumer Price Index
CPL - Capillary Pumped Loop
CPU - Central Processing Unit
CRT - Cathode Ray Tube
CSFL - Continued Safe Flight and Landing
CTE - Coefficient of Thermal Expansion
CWT - Center Wing Tank
DARPA - Defense Advanced Research Projects Agency
DC - Direct current
DER - Designated Engineering Representative
DEW - Directed Energy Weapon
DNI - Direct Normal Irradiance
DOD - Department of Defense
DOY - Day of Year
EASA - European Union Aviation Safety Agency
ECS - Environmental Control Systems
EE - Electrical/Electronic
EIDI - Electro-Impulse Deicing
EMF - Expanded Metal Foil
EMI - Electromagnetic Interference
EOL - End of Life
EoT - Equation of Time
FAA - Federal Aviation Administration
FCAC - Forward Cargo Air Conditioning
FDS - Fire Dynamics Simulator
FLUINT - FLUid INTegrator
FQIS - Fuel Quantity Indicator System
FSU - Former Soviet Union
GAO - Government Accountability Office
GHI - Global Horizontal Irradiance
GMT - Greenwich Mean Time
GPS - Global Positioning System
GUI - Graphical User Interface
HEPA - High-Efficiency Particulate Arrestor
HRR - Heat Release Rate
HX - Heat Exchanger
IATA - International Air Transport Association
ICAO - International Civil Aviation Organization
ICBM - Intercontinental Ballistic Missile
IFE - In-Flight Entertainment
IR - Infrared
ISA - International Standard Atmosphere
ISO - International Organization for Standardization
ISS - International Space Station
IT - Information Technology
JFK - John F. Kennedy
LAR - Live Animal Regulations
LASER - Light Amplification by Stimulated Emission of Radiation

LCD - Liquid Crystal Display
LED - Light-Emitting Diode
LH2 - Liquid Hydrogen
LOX - Liquid Oxygen
LSL - Low Solar Load
LST - Local Solar Time
LSTM - Local Standard Time Meridian
LT - Local Time
MDO - Multidisciplinary Design Optimization
MEA - More Electric Aircraft
MET - Metabolic Equivalent of Task
MIL-HDBK - Military Handbook
MIL-STD - Military Standard
MLG - Main Landing Gear
MPS - Minimum Performance Standards
MRT - Mean Radiant Temperature
NACA - National Advisory Committee for Aeronautics
NAMS - Nautical Air Miles
NASA - National Aeronautics and Space Administration
OEV - Overboard exhaust valve
OFV - Outflow valve
UV - Ultraviolet

Mark Ahlers is a retired Technical Fellow in thermal and fluid flow analytical tool and process development to support vehicle system integration at the Boeing Commercial Airplane group. He spent 30 years developing the analytical tools and processes which form the basis for aircraft thermal management on Boeing commercial airplane programs, in addition to supporting thermal design efforts for the International Space Station, launch vehicle proposals, commercial and research satellite, and military derivatives of commercial aircraft programs. When thermal issues were recognized as a serious threat to the economic implementation of a more electric systems architecture and more temperature-sensitive composite structure on the 787 program, Mark was given responsibility for technical oversight of thermal activities as Boeing's first Thermal Marshal.

Mark has taught hundreds of aerospace industry employees of all disciplines the theory, history, and real-world application of ATM shared in *An Introduction to Aircraft Thermal Management*. He has a BS degree in Mechanical Engineering from Colorado State University, and is the author of *Aircraft Thermal Management: Integrated Energy Systems Analysis*, and *Aircraft Thermal Management Systems Architecture*, which are both published by SAE International.

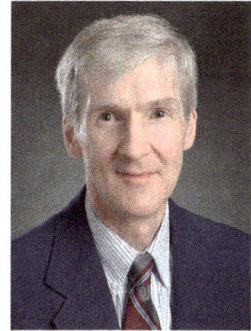

index